THE FUTURE OF
WAR

THE FACE OF 21ST-CENTURY WARFARE

MARC CERASINI

ALPHA

A Pearson Education Company

The future's so bright, I gotta wear shades ...

—Timbuk 3

CONTENTS

FOREWORD

Marc Cerasini opens this book with the quote, "The future's so bright, I gotta wear shades." The future of warfare is indeed bright, and the technology that makes it so is advancing at an astonishing pace.

Advances in warfare have always been limited by available technology. Until the development of the combustion engine, the maximum speed of movement to—and on—the battlefield depended on the speed of man afoot or on horses, elephants, camels, and mules. In the near future, so-called "scramjets" will cruise at 8.5 miles per second, carrying weapons across the continental United States in 6 minutes flat.

Up through the early 1900s, a commander's best source of battlefield intelligence (who or what is on the other side of the hill) depended on horse-mounted cavalry penetrating into enemy territory, locating the opposing forces, and trotting back to report on the enemy's whereabouts and activities. Marc describes sensors that detect the enemy even if he's lurking around corners or sneaking through the heaviest of fogs or densest of night skies. And of course, ever-increasing advances in satellite reconnaissance technology make it possible to read the writing on an enemy vehicle from space.

My, how times have changed.

But the advances just mentioned rely on *existing* technology. Marc probes further into the future, giving us a glimpse of what researchers are dreaming up in their Department of Defense–funded think tanks and university research programs: exoskeletons that will turn men into virtual machines, battle uniforms that will diagnose and treat injuries, electromagnetic rail guns that will propel jets from carriers, "bullets" that will eat polymers.

To turn these fantasies into reality, research is ongoing in fields as far flung as miniaturization, nanotechnology, materials science, and medicine, to name just a few.

Thankfully, Marc adds a dose of reality to these imaginings, touching on the same concerns I worried about in my final Air Force assignment as Chief of Staff of the U.S. Air Force Space Division. We routinely spent $4 to $6 billion per year in developing space-related technologies and buying advanced satellites and launch vehicles. With these kinds of price tags, if you don't worry about cost vs. quantity, reliability, and vulnerability, you've got your head in the clouds. (And did I mention data-overload? September 11 taught us something about that.)

Hi-tech weapons have high-cost price tags. Sure, having the most advanced equipment can win wars; but if you can only afford one or two key weapon systems, you run great risks. America built nearly 13,000 B-17s, one of the workhorse bombers of World War II. In 2002, the Air Force has barely 1 percent of that number of bombers—counting B-1s, B-2s, and B-52s. We need breakthroughs in costs similar to those that made digital clocks so cheap that manufacturers could easily afford to add them to almost anything. When a single satellite or bomber costs $400 million, you need reliability. Most satellites can't be fixed in space. Nevertheless, the Air Force has built in remarkable reliability on many military satellites placed in orbits that cannot be reached by the *Space Shuttle*. The signals from 24 GPS satellites provide precision worldwide navigation for American forces. Yet if the GPS constellation becomes vulnerable, I can only wonder if any backup navigational equipment will exist and whether anyone will know how to navigate without the marvels of GPS.

In this book, Marc reviews upcoming improvements for all services as well as some far-out, imagination-defying proposals. His revelations will whet your appetite to learn more, and he provides an extensive appendix of references for future study.

So put those shades on and turn the page.

Col. Jimmie Butler, USAF (Ret.), and author of *A Certain Brotherhood*

PROLOGUE

Judging by the tide of human events at the beginning of the 21st century, it is clear that humankind is on the verge of a massive technological, sociological, and cultural shift that will forever change the way we live. Further scientific and technological progress is inevitable. Current and future advances will continue to reshape our world—a world that is shrinking every day. Global communications, travel, and trade both unite and divide cultures and nation-states. Within the complex and shifting societies of the contemporary world, tension is bound to erupt.

Throughout human history, governments, cultures, and religions have clashed—sometimes violently. There is no reason to believe that this situation will change in the near future, so it is the duty of all democratic nation-states to maintain the peace and protect their citizens, even as they prepare for war. Sadly, preparing for war has become a full-time job in the modern world. Only a vigilant, well-equipped, professional standing army can guarantee national security. The tempo of war has so accelerated in modern times that any conflict would be over long before a conscripted army could be assembled, trained, equipped, and fielded. There is no longer sufficient time for a nation-state to "gear-up" and "mobilize" as America did at the start of World War II.

The same revolutionary scientific advances that have reshaped our society will also revolutionize the way humans wage war. Though the coming military revolution is still distant, vague, and undefined, such a transformation is inevitable. In a sense, all modern warfare is a struggle between weapons systems. Technology is used to improve competing systems, but, at some point, one weapon or tactic achieves an overwhelming advantage over its rival, and the war is over. Though there is no way to predict what types of "ultimate weapons" will result from the constant competition between weapons systems, we must hope that our military leaders are wise enough to recognize *useful* change, and are open and ready to embrace and adapt to such change. Otherwise, national security may be jeopardized.

The seeds of the radical battlefield revolution to come have already been planted. Some have begun to bear fruit, and now that the genie is out of the bottle, no power can prevent the impact of science and technology on the military arts. I believe it is our duty, as citizens of the most technologically advanced and economically powerful democratic nation on Earth, to attempt to familiarize ourselves with the transforming technologies looming just over the horizon. One day—perhaps sooner than we think—new weapons designed for the "next great war" will be deployed to protect us from our enemies.

As American citizens, we pay for these weapons, and they will ultimately be used in our name, to protect our Constitution. The least we can do is know what these expensive and deadly systems are capable of doing. And as citizens of a free society, we should also become well versed in the theories, strategies, and tactics that will guide the deployment of these weapons, so that we can make an informed judgement when it comes time to state our opinion, or cast our ballot.

ACKNOWLEDGMENTS

The author would like to thank Colonel Jimmie Butler, USAF (Ret.), for his invaluable effort. I would also like to thank my supervising editor at Alpha Books, Gary Goldstein, for his dedication, patience, and persistence. I must also express my gratitude to Jennifer Moore for the remarkable feat of reshaping a too-long, overly technical manuscript into the work you now hold in your hand.

Many people in the defense community also gave their time and effort to the creation of this book. I would like to thank the folks on the F-22 Raptor Team at Lockheed Martin, Boeing, Pratt & Whitney, and the U.S. Air Force for their assistance. I would also like to extend my gratitude to the personnel at General Dynamics Land Systems, Northrop Grumman, and the Oak Ridge National Laboratories for their tireless support. Any mistakes that made it to the final draft are solely the responsibility of the author.

PART 1

THE PAST INFORMS THE FUTURE

"War is being drawn into the field of the exact sciences. Every additional weapon, every new complication of the art of war, intensifies the need of deliberate preparation, and darkens the outlook of a nation of amateurs. Warfare in the future, on sea or land alike, will be much more one-sided than it has ever been in the past, much more of a foregone conclusion."
—H. G. Wells, *Anticipations*, 1902

"… science, being universal and neutral, knowing neither good nor evil, offered its gifts for men to use them as they wished. But once technology had been committed to the conflict between nations, it was inevitable that one day war would become fully automatic."
—I. F. Clark, *Voices Prophesying War*, 1992

"The path of progress is strewn with the wreck of nations."
—Karl Pearson

CHAPTER 1

THE WAY OF THE GUN: OBSOLESCENCE?

History has taught us many lessons, the most important of which is that wars are inevitable. In this century, as in the last, international conflict—or the threat of conflict—will continue to dominate and define the way nation-states relate to one another. Wars will continue to be waged, whether the United States is involved in these conflicts or not. And as our economic and political influence increases, so, too, do the potential political, economic, and military threats arrayed against us. It is increasingly apparent that Americans no longer have the luxury of "sitting out" of what appear to be small, isolated wars, even if they are fought on the far side of the globe, and even if they do not always appear to pose a "clear and present danger" to the well-being of the nation. In a world of instant global communications and international commerce, there is no such thing as an "indirect threat"—any armed conflict has the potential to harm an ally, cripple international trade, or halt the flow of natural resources. Any such action would surely impact the "American way of life," regardless of the time and distance separating "us" from "them."

The way in which wars are fought is poised to undergo a fundamental transformation as a result of the wealth and technology of the modern nation-state. The United States is the world economic and technological leader, and shall remain so for the foreseeable future. During the 20th century, we saw the seat of world power shift away from Europe and Asia and toward North America. This shift will likely continue through the 21st century, making the United States and its North American neighbors the focus of political and economic power for some time to come.

Much of America's power and influence is the result of its successful deployment of its formidable military might. American blood, sweat, and industrial efficacy defeated fascism in Europe and thwarted Japan's imperial ambitions during World War II. America's ability to design, build, and deploy advanced, highly effective (and wildly expensive) weapons systems contributed to the collapse of the former Soviet Union in the 1990s. America's technological advances and formidable military power will assure the continued preeminence of the United States in the 21st century.

At the center of this military might is a basic invention that first emerged in Western Europe more than five centuries ago. Despite the array of complex killing technology displayed on the modern battlefield, the most basic weapon of war is still a chemically propelled object fired through a hardened metal tube: technological innovations we call the *gun* and the *bullet*. Though guns today are more powerful and more accurate, possessing a greater range and a higher rate of fire, this weapon system has not changed in its basic concept since it first appeared in the 14th century. In all projectile weapons—from an ancient blunderbuss to a modern artillery piece—a chemical reaction causes an explosion within a tube, forcing out a hard object with enough velocity to propel it forward and inflict damage on an enemy soldier, emplacement, or vehicle.

Guns still rule, and will probably dominate warfare for many decades to come. Indeed, since the first primitive muskets and flintlocks were fired, virtually all innovations in warfare have been made primarily to make the gun more *powerful*, more *accurate*, and more *readily available* to warriors on the battlefield. Look at the evolution of the armored tank: Beyond the addition of a large, petroleum-fueled engine, a pair of caterpillar tracks that can negotiate most terrain, and impenetrable steel or ceramic armor, the armored tank is nothing more than a *weapons platform* built to carry a cannon and a few machine guns (projectile weapons all) close enough to the enemy to inflict damage—and hopefully survive to fight another day. The tracks, armor, and engine of a tank are really of secondary importance—they exist solely to bring the gun within killing range of the enemy. The same is true of a modern, ocean-going destroyer, a fixed- or rotary-wing aircraft, or a jeep with a bazooka mounted on its roof. Like the phalanx, the chariot, the siege engine, medieval armor, and the Norman longbow, the projectile weapon was a *basic innovation* that altered the way wars are fought. The platforms that haul the gun into battle are secondary innovations that have changed considerably in the 20th century, but the basic projectile weapon has not.

The gun has constantly evolved over its long history, piling modification onto modification until the projectile weapons' development reached critical mass in the 20th century. From black powder, to breech-loading and rifled barrels, to smokeless powder, and on to automatic weapons and explosive cartridges—each stage in the gun's evolution was an effort to maximize the weapon's effectiveness in battle. As with all weapons, the gun began its life on the edge of the battlefield. Early firearms shared the stage with swords, crossbows, and armor, and were deemed inferior to all of them. The first guns were such primitive instruments that it took two centuries for them to find military acceptance, and even more time to move from the periphery to the center of battle.

When they first appeared in the 14th and 15th centuries, muskets and blunderbusses were—for a variety of reasons—quite unreliable. Metallurgy was not yet an exact science, so the barrels of early guns were prone to explode with continued use. During this period of birthing, projectiles were chosen more for their availability than their aerodynamic qualities. But neither rocks, nor buckshot, nor formed ball-shaped pellets fly on a straight-and-true trajectory, so the chance of actually hitting an enemy soldier with an early musket or flintlock was less than optimal. Added to that, the gunpowder was often impure, resulting in yet more variations in performance.

But an even bigger obstacle had to be overcome before the gun became the central weapon of war. The rate of fire with a primitive musket, blunderbuss, or flintlock was quite slow. It could take many seconds—even a minute or more—just to reload. In warfare, a lot can happen in a minute. Indeed, even a few seconds can seem like an eternity in battle. To counter this damaging limitation, guns came to be used in mass attacks in an effort to increase their effectiveness. The logic was simple. A whole bunch of badly aimed shots fired through poorly made metal barrels had a much better likelihood of doing damage to the enemy than a few shooters firing sporadically. And if an army's musketeers were divided into teams, so that one team did the firing while the other team reloaded, a massed attack had an even better chance of repulsing the enemy. Accuracy was not as important as numbers—all that mattered was that there were a lot of guns firing in the general direction of the opposing army at approximately the same time.

As projectile weapons increased in power they also increased in size. Alongside handheld weapons there evolved large, wheel-mounted, horse-drawn guns with heavier barrels and greater destructive power. So it was that projectile weapons like cannon and artillery were born—adding to the lethality of the battlefield. Soon these artillery and explosive shells replaced outmoded siege engines. The invention of gunpowder and the gun finally overwhelmed the stationary, fortified strongpoint—a castle or walled city—and rendered such structures forever obsolete.

The use of artillery and massed fire reached its apex with the deployment of the machine gun on the Western Front of World War I. Tragically, it took time for military leaders to realize that in a modern, mechanized war, victory belonged to the army that could fire the most projectile weapons in the shortest period of time. In the beginning of that war, military tactics had not yet caught up with technology, so the British, French, German, and even American armies all charged into the teeth of entrenched machine guns at various times during the protracted and bloody conflict—and all suffered horrendous casualties. It was on the Western Front of 1914 that outmoded 19th-century battle tactics collided headlong with the mechanized death that would become the future face of war.

As with the gun itself, the platform that carries the projectile weapon into battle has evolved. In the 60 years between the American Civil War and the outbreak of World War I, only *three principal* weapons platforms have emerged: the battleship, the tank, and the bomber.

The first to arrive in the early part of the 20th century was the sea-going *battleship*. The battleship evolved through the combination of two previous inventions—the wooden, wind-powered European gunships used for exploration and colonization during the 15th, 16th, and 17th centuries, and the Union and Confederate ironclads that fought on the rivers and along the coastal regions during the American Civil War.

Like the sail-powered wooden gunships, the battleship was created to project European military power to far-away regions and to defend colonial interests. Like the ironclads, battleships were constructed of metal and heavily armored, so they could destroy the enemy without being destroyed. 20th-century battleships were equipped with powerful oil-and-steam engines and large, powerful cannons that could pour a barrage of explosive projectiles onto another warship or enemy shore positions. With the arrival of more efficient petroleum-fueled engines, battleships increased in size and power, reaching their zenith with the dreadnoughts floated by the European powers during World War I.

The British response to the devastation wrought by the machine gun in World War I was the armored tank, the next platform innovation of 20th-century warfare. The tank is an armored mobile platform designed to carry a cannon or brace of machine guns into the enemy ranks while remaining impervious—or seemingly impervious—to assault. Powered by an internal combustion engine, the tank was designed by the British as an anti-infantry weapon during World War I. Armored juggernauts like the gigantic British Mark I could also be used to ferry troops safely to the battlefield behind the tank's protective armor plating. In common with the battleship, the tank was equipped with one or more cannons, and/or a brace of machine guns used to clear trenches and break through enemy lines. When this weapon first appeared on the battlefield, the psychological effect on German soldiers was devastating.

The tank appeared at first glance to be a totally new, even revolutionary development—a weapon both formidable and indestructible. Yet looks can be deceiving; the concepts behind the tank were far from new. In truth, the tank was an ingenious combination of a 10th-century fortification (albeit a mobile fortification), an 18th-century ironclad (albeit one that moved over land instead of water), and a 20th-century internal combustion engine—unified for a single purpose: to carry troops and projectile weapons closer to the enemy to inflict the maximum amount of casualties.

Like the armored tank, the addition of the airplane to the arsenal of war appears at first glance to be a principle innovation, but it is not. Why? Because the airplane—like the tank and the gunboat—is simply a *platform* to carry projectile weapons to the enemy. Even the air-fired, rocket-propelled missile—a technological innovation that arrived in World War I and was improved during World War II—is nothing more than a projectile weapon that replaces the directing metal tube with a rocket engine and fins or wings that guide the missile to its target. Like the tank, the airplane is nothing more than a means to engage the enemy with *projectile weapons*.

The final weapons platform to emerge in the 20th century was the *airborne bomber*. Though aerial platforms such as hot air balloons had been used for reconnaissance in Europe as early as 1843 and two decades later during the American Civil War, the concept of aerial warfare didn't come into its own until the development of the bomber during World War I. Powered by petroleum-fueled internal-combustion engines, the bomber was designed to deliver explosive projectiles onto the enemy in the form of bombs. Though these bombs were dropped instead of fired from a tube, bombers were still forced to utilize projectile weapons in order to protect themselves. The fighter plane later emerged—primarily as an airborne weapons platform designed to protect the bombers. The fighter's evolution into a reconnaissance platform and a bomber in its own right would come later.

All weapons platforms have a life cycle that begins with creation, continues with a rise to dominance, and ends in obsolescence. These three—in their many varieties—dominated warfare from the beginning of World War I to the outbreak of World War II. The tank and the bomber are still in use today, but the battleship has been superceded by the aircraft carrier.

The end of the battleship's domination seemed to come suddenly, unforeseen by all but the most visionary naval strategists. Yet the seeds of the battleship's obsolescence were sown much earlier than its demise.

In 1940, the fragile British biplanes sent to torpedo the German battle-ship *Bismarck* appeared, to one observer, "like fleas attacking a bear." Yet, those "fleas" were able to inflict enough damage to the *Bismarck*'s rudder to slow it down until British warships arrived on the scene to finish her off. At Pearl Harbor, Japanese carrier-launched aircraft sank the bulk of the U.S. Navy's fleet—including the "invincible" battleships. Later, at the Battle of Midway, the final nail in the battleship's coffin was hammered home when U.S. carrier-launched dive bombers and torpedo planes sank the bulk of the Japanese navy, without American warships getting close enough to fire a shot. The Imperial Japanese navy could never mount an offensive again, and resorted in desperation on a defensive posture that all but doomed Japan's ambitions.

By the end of 1942, the centrality of the battleship in naval operations had been permanently eclipsed by the aircraft carrier, and the outmoded battleship task force was quickly transformed into the *carrier* task force— the aircraft carrier and its medley of supporting vessels—that prowls the oceans today.

This cycle of innovation, dominance, and obsolescence seems to be immutable. If the once-mighty battleship could suddenly be rendered obsolete, can armored tanks and airborne bombers be far behind? And if tanks and bombers fall into decline and disuse, could not the same thing happen to the now-dominant projectile weapon?

The gun is such a basic component in modern warfare that for many—including many members of the present generation of military strategists—it is hard to believe that the Era of the Gun may soon end. So far, the evolution of the gun has been spectacular, but like the char-iot, medieval armor, and the longbow, the days of the gun's dominance on the battlefield are numbered. That is not to suggest that the gun, nor the concept of the projectile weapon, will become obsolete overnight. The gun, in its many and varied forms, will most likely exist well into the 21st century, but probably not beyond it. Within the last 30 years, a new type of weapon has emerged, a weapon so simple yet so revolu-tionary that it will surely spell the end of the projectile weapon's domi-nance within this century.

The projectile weapon has one major weakness that will ultimately cause its downfall. Once a projectile weapon is aimed and fired, the laws of physics take over. If the person who fired the weapon is a good marksman, then the bullet or shell has no better or no worse than a 50/50 chance of hitting the target. Actually striking the target will depend on variables such as distance, wind velocity, temperature, and so on—all out of the shooter's control. That is why projectile weapons are deployed en mass—more bullets and shells mean more hits.

But as we saw during the Persian Gulf War and the war in Afghanistan, the United States now possesses the first generation of so-called "smart weapons," which use computers and microchips to *guide themselves* to the target. *Precision-guided munitions* can reverse those laws of physics, giving control of the projectile back to the shooter, even after it has been fired. The visible results of the deployment of this gigantic leap in technology were broadcast around the world during the Gulf War. As dramatic as televised shots of a single guided bomb destroying a bridge, bunker, or aircraft were—bear in mind that those results were achieved by *first generation technology*—they will only improve. In fact, the new generation of precision-guided weapons (PGMs) is already smaller, faster, more accurate, and more deadly. It is hard to believe that this technology has been around for fewer than 60 years. Strategists are already predicting that PGMs as small as bullets—small enough to be fired from a sidearm—are only a few decades away. And that is only the tip of the technological iceberg.

Indeed, the future battlefield will likely be filled with scientific wonders hardly envisioned now. We already possess infra-red viewers that can pierce the darkness; bombers that can be rendered invisible to radar; munitions that guide themselves to the target; satellites that can detect enemy vehicles and emplacements—even if they are buried under tons of earth and concrete; instant communications with any point on the globe; unmanned, remote-controlled reconnaissance aircraft that can hover for hours, days, or even weeks over the battlefield and even launch a missile attack; and electronic weapons that can jam enemy radar and radios, and perhaps even destroy the microchips inside their battle-management and targeting computers.

THE FIRST PRECISION-GUIDED WEAPONS

Precision-guided bombs made their debut in World War II, a fact that is curiously ignored by historians. Germany developed two types: the Ruhrstahl/Kramer X1—or simply the *Fritz X*—and the more advanced Henschel Hs 293. The Fritz X was a guided glide bomb. The Hs 293 was powered, using a rocket to maneuver into glide position. Both used radio waves to guide the bomb to the target.

First use of precision guidance was on August 27, 1943, when a German Hs 293 sank the HMS *Egret*. But the Fritz X had a more spectacular debut against the Italian battleships *Italia* and *Roma*. After the collapse of Mussolini's government, Hitler feared the Italians would turn Italy's warships over to the Allies. To prevent this catastrophe, the Luftwaffa dropped two Fritz X bombs on the battleships from 18,000 feet. Using line-of-sight vision, the bombardiers manipulated a joystick, which generated a radio signal that dispatched commands to the bombs' receivers. The bombs fell for a minute before they struck. The *Italia* was heavily damaged, and the *Roma* was destroyed when the bomb struck her main magazine and blew the battleship in half, killing a thousand of her crew.

In the not-so-distant future, armies will deploy chemicals that solidify the fuel inside the tanks of enemy vehicles; bacteria that can "eat" rubber tires; mines that can recognize an enemy vehicle and destroy it; weapons systems with enough artificial intelligence to fire themselves; cyberuniforms embedded with computer chips with the ability to monitor body functions, diagnose a soldier's wounds, administer an anesthetic, and begin triage; stealth suits that can change color to suit their environment; guns that can fire around corners; and an arsenal of nonlethal weapons designed not to kill enemy soldiers but to "neutralize" them.

Such technology will almost certainly result in a transformation of the world's military forces. No longer will nations strive to build mass armies made up of thousands of armed troops, in the hope of overwhelming its opponent, because instead of sheer numbers, the mastery of complex technology will be the key to victory. Such a sea change may also result in dramatic social and political reorganization in the not-too-distant future, because any advanced industrial nation—no matter its size or population—could conceivably create and deploy a technologically advanced military force with the ability to defeat a numerically superior army through the use of high-tech, cutting-edge weapons alone.

CHAPTER 2

A COUNTRY MADE BY WAR

Everyone—from leaders of industrialized nations to third-rate dictators and power-hungry terrorists—knows that the United States has the most potent military in the world. But how did America come to possess its formidable military might? Why was such a military not created by more belligerent nations like the Soviet Union, or Nazi Germany, or Imperial Japan? Why did the United States—ostensibly a peaceful nation with no aggressive design on its neighbors—end up as the sole superpower at the dawn of the 21st century?

The evolution of military power and military technology made great leaps in the United States during the 19th century, although the stage was set for America's military ascendance as far back as the American Revolution. Indeed, only the United States could have achieved such military power, because the United States and *only* the United States possessed the unique political, cultural, numeric (population), geographic, economic, and scientific conditions that are conducive for the creation, maintenance, and support of a large, technologically superior military like the one we have today.

The political and cultural stage was set for America's military dominance in the 21st century back in the earliest days of the colonial period. In the 1770s, the American colonies were in revolt against the British crown and the formidable power of the British military—then the greatest war machine on Earth. In the 18th century, as in the centuries before, European kings and princes made war using large standing armies comprised of professional soldiers. The individual within those armies possessed almost no human rights—they were disciplined harshly, ordered to move or fight without regard to their own wishes, and could be used—or squandered—in any manner the kings or princes who held absolute sway over them saw fit.

Military organizations evolved a little differently in the New World. When the first European settlers arrived, they brought guns, powder, and cannons with them, but few professional soldiers. Defense of settlements and communities was the sole responsibility of the members of those communities. Local militias were formed to meet any challenge, from Indian incursions to bandits to threats from hostile European powers projected from far across the Atlantic. So it was that from the beginning of the American Revolution, even before the Declaration of Independence or the ratification of the Constitution, the concept of the "citizen-soldier" gained widespread prominence and became widely associated with the colonial militia. This concept is based on the notion that citizens have the right and the obligation to arm themselves to defend their community or nation from enemies foreign or domestic, and it runs counter to the other forms the military took in 18th-century Europe.

While the European states continued to make war using large, professional standing armies comprised of soldiers with few rights, in the New World a new sense of community responsibility was born. While the soldiers in European armies were isolated from the society they were compelled to defend—and could potentially pose a threat to the legitimate government—militias in America were comprised of members of the community rallying to a common cause.

George Washington's Continental Army was a *volunteer* army, and, as such, the individual soldier was regarded as a free citizen born with certain inalienable rights. These rights were not granted to individuals from the king—as in the Magna Carta—nor did they descend from any human institution. Rather, these rights were "self-evident" and endowed to American citizens at birth, by "their creator"—thus, no human government or hand had the right to curtail these rights.

These fundamental human rights created two quandaries for the nascent American republic. The first was a basic distrust of a standing army and professional soldiers, both of which were perceived by citizens of the new American republic as corrupt tools of despotism. The result of this innate distrust was that, despite the heroic performance of the Continental Army, once the War for Independence had ended and the American republic was established, that standing army was all but dismantled, its officers cashiered out.

The other quandary posed by democratic human rights proved more difficult for military leaders. Citizen-soldiers united under one flag for the cause of mutual defense could not be used in the traditional manner of European armies. These men were not property, but free citizens of the United States. As such, their lives were precious and would never be wasted. American soldiers were not "cannon fodder." Political and military leaders who needlessly endangered the men under their command faced losing their command in the next election or military review. Massive casualties became unacceptable, and U.S. military leaders were faced with the problem of avoiding undue hardship and suffering among their troops.

Another problem faced by America's military leaders was numeric—that is, the relatively small size of the American population when compared to the population densities in the rest of the industrialized world. America has a low population when compared to its land mass and its defensive responsibilities in the post–Cold War era. The population density of North America is a mere 68 people per square mile. Compare that with France's 252, Germany's 400, or Japan's 844 people per square mile.

Historically, the manpower and intensive organization of European armies imposed a considerable strain on the limited population in America. This trend continued through World War I, World War II, Korea, Vietnam, and the Persian Gulf War. In each conflict, American military forces were much smaller than their enemy's. Inferiority in numbers compelled U.S. military leaders to find a way to counter the dangerous fact that the U.S. military would be outnumbered in almost any conflict.

Geographically, the United States is one of only two nations that possesses large coastal regions facing both the Atlantic and Pacific Oceans (the other nation is Canada). Areas of control seem to expand with civilization. While ancient Greece and Rome needed only to control the Mediterranean Sea to assert their military and political dominance, Great Britain was compelled to control the entire expanse of the Atlantic Ocean in the centuries leading up to the Second World War. In the first half of the 20th century, Japan sought dominance of the Pacific Rim as a way to project its growing political and economic clout. Japan did not seek to expand its influence into the Atlantic Rim, because it had no interest there.

The United States of America has economic and strategic interests in both oceanic regions. Today—as it has been since the transcontinental railroad permanently connected America's east and west coasts—the United States is compelled by its own unique geography to project naval power and military force in both the Atlantic and Pacific oceans. That it is capable of projecting such might makes the United States the first-ever pan-oceanic superpower in human history, but this distinction comes at great cost, for maintaining a military presence in both oceans increases manpower demand, and the economic cost stretches resources even further.

Of course, the advantages of having busy ports in both oceans far outweigh the disadvantages. Possessing lucrative trade routes in both oceans contributed mightily to the economic dominance of the United States, and immigration from societies in both the Atlantic and Pacific Rims enriched our culture even as trade enriched the treasury.

Economically, the United States is likely to remain the powerhouse of the world well into the future. The creation of wealth is an intrinsic part of the American psyche—the American dream. The infrastructure of the United States, the rapid pace of innovation, and the advances of science concentrated within America's borders is unrivaled. American industrial productivity is the envy of the world, and likely to remain so for decades to come. Although other economic powerhouses have risen and fallen—especially in the 20th century in the Pacific Rim—the health and stamina of the U.S. economic machine has remained more or less constant, even in the face of naturally occurring economic cycles, war, and natural disasters.

The rapid pace of innovation and scientific advancement sets the United States apart from every other nation. Since the birth of the republic, Americans have turned to science and technology to solve problems and to ease the burden of day-to-day living. It was science and technology—in the form of railroads—that united east and west. And it was technology that Americans turned to in times of great crisis. So, it was natural that Americans should turn their dominance in science and technology to the arts of war.

From its birth, America depended more on the *quality* of its warriors and weapons than on its numbers. Though the Continental Army was sorely outnumbered by the British, they were fighting on their own turf (an advantage that cannot be overlooked) and the American soldiers were infinitely better marksmen than their British counterparts.

AMERICA TAKES AIM

It was the American army that added the word "aim" to the phrase "Ready. Aim. Fire!" That is because trained woodsmen, trappers, and Indian fighters understood the value of aiming their weapon before pulling the trigger. By doing this, the soldiers of the Continental Army were utilizing the concept of firepower, applying the forces they have more accurately, in order to balance out their opponent's superiority in numbers.

During the American Civil War, North and South vied for technical dominance as much as military superiority—especially on the high seas. Although the South won most of the battles, they were eventually defeated by the superior industrial capacity of the North, and the federal government's ability to procure better, more advanced weapons and deploy them faster. Throughout this conflict, both sides improved the quality of their weapons, but the North did it more efficiently. The vision of Eli Whitney was finally fulfilled in Union arsenals like the one in Springfield, Massachusetts. Using interchangeable parts and mass-production methods promised by Whitney 40 years earlier, this single facility churned out 800,000 guns between 1860 and 1865.

By the time of America's involvement in World War I, the United States had lost much of its defensive infrastructure and capacity. U.S. troops were forced to rely on often-inferior weapons of European manufacture. By the Second World War, America had recreated its industrial might, but was still compensating for its inferior manpower— this time by building vast quantities of somewhat inferior weapons. The Sherman tank, for instance, was inferior to the German Tigers and Panthers they faced. But the United States produced so many Shermans that they overwhelmed the nominally superior but numerically inferior Tigers and Panthers.

Only in the air did American designers and manufacturers excel, producing superior machines that outpaced the aircraft fielded by both America's allies and opponents. Aircraft like the P-47 Thunderbolt and the P-51 Mustang were, respectively, the greatest attack planes and fighter aircraft of their era. Even then, America was lucky, for Germany *did* manage to design and create superior products like the V-1 and V-2 rockets, and the world's first operational jet fighter, the Messerschmitt 262—just not in sufficient numbers, and far too late to make a difference.

From the 1950s to the present, America continually improved the quality of its weapons—making do with fewer and fewer weapons, but building much better ones. American military planners kept pushing technology to the limit. In the end, they managed to produce advanced, high-tech weaponry like nuclear submarines, aircraft carriers, tactical fighters, attack planes, and strategic bombers that were larger, faster, and better than their rivals. In the 1970s, America began to deploy precision-guided munitions and smart bombs, further upping the ante. With rockets that fly to their own target and hit it precisely, and advanced designs like the Nimitz-class carrier, the F-22 Raptor, the AH-64 Apache attack helicopter, and the Los Angeles–class submarine, America's military assets have so outperformed the products of other nations that the

United States has become the unrivaled martial power of the world simply because *we own them*.

CHECK, PLEASE ...

According to the *CIA World Factbook, 2000,* in 1999 the United States spent $200 billion more on defense than any other country. The United States spent $276 billion; Russia came in second at $56 billion.

Of course, the United States paid a premium price for these weapons (that's where being an economic powerhouse comes in!), but what America got for its money was high-performance, top-of-the-line weapons systems and platforms that provide more bang for the buck than anything made anywhere else—a fact that was vividly made apparent by the performance of the U.S. military in the Persian Gulf War.

CHAPTER 3

THE FUTURE HAS ARRIVED: THE GULF WAR

On the early morning hours of January 17, 1991, hundreds of Tomahawk cruise missiles blasted from the decks of a dozen U.S. Navy warships. These ships were moored in the Persian Gulf or the Gulf of Oman—literally hundreds of miles from the final destination of the Tomahawk swarm that blasted out of their launchers. Streaking on fiery tails into the night sky, the missiles were bound for military and strategic targets deep inside Iraq.

Tomahawk Land Attack Missiles (TLAMs) are among the most advanced cruise missiles of our time. At a relatively frugal cost of a few million dollars a unit, these computer-guided, long-range missiles are designed to attack strategic ground-based targets, and provide a big bang for the bucks. Fired from a shipboard tube, the Tomahawk has a torpedo-shaped fuselage and looks like a rocket when first launched. But within moments, a ventral turbo-fan intake, control wings, and tail surfaces are deployed from inside the hull. Now transformed from a rocket into a rocket-powered aircraft, the Tomahawk suddenly resembles its distant, primitive cousin—the sleek, unguided V-1 rocket planes that Hitler fired at London in the closing months of World War II.

At approximately 18 feet long, with a wingspan a little less than 9 feet, each Gulf War-era TLAM easily achieves a range of 800 miles while traveling at an average speed of 450 miles per hour. The Tomahawk cruises at an altitude of between 50 and 100 feet—well below the defensive radar capabilities of most nations. Controlled by two separate on-board guidance systems, navigated by a GPS (Global Positioning System) receiver, directed by Honeywell AN/APN-194 short-pulse radar, using a Maverick Imaging Infrared seeker to pierce clouds, fog, and darkness, all these complex systems inside the Tomahawk are co-ordinated by a powerful, on-board IBM digital computer brain.

Once fired from their gray steel launch pads that January morning, the Tomahawks raced over the waters of the Gulf, to reach landfall in a remote area of the Middle Eastern coast. Flying the nap of the earth, the cruise missiles plunged into the rugged mountains of western Iran. While traveling through mountainous landscape, Tomahawks utilize advanced terrain-contour-matching techniques—that is, the Tomahawk uses its on-board radar altimeter to compare, millisecond by millisecond, the mountains, valleys, bluffs, and dry river beds it flies over with a digitized map drawn from satellite images and stored inside its electronic brain.

Following their pre-programmed route, the Tomahawks dropped down over the foothills of the Iranian mountains, and then veered left to cross the border of Iraq. At that point, the TLAMs scattered further, each to strike at its predetermined target in and around the Iraqi capital of Baghdad. Military installations, arsenals, storage areas, military vehicle depots, radar sites, government buildings, communications centers, and power plants were all considered legitimate targets.

Within 10 miles of their pre-arranged destinations, the Tomahawk shifted navigational systems, switching to an optical sensor-system called a digital scene-matching area correlator (DSMAC). Using DSMAC, a television camera mounted on the fuselage scans the passing scenery and instantaneously compares what it "sees" to images stored in the on-board computer. Inside the brain of each Tomahawk, guidance computers issue final commands—adjusting the trim, speed, and angle of attack using the stubby wings and tail surfaces.

Late-night eyewitnesses on the streets of Baghdad who managed to catch a brief glimpse of the Tomahawks, or heard the missiles streak overhead, usually detected the TLAMs at this point in their trajectory— when the cruise missiles slowed to turn corners or pop high into the air to hurl themselves over obstacles. All this happened in the brief moments before the Tomahawks zeroed in on their targets and made the kill.

At approximately 3:05 A.M. Baghdad time, the first Tomahawks arrived. Already, the skies over the Iraqi capital were ablaze with anti-aircraft guns. But every gun was firing blindly, filling the night with bright yellow tracers. Cruise missiles slipped between the ineffectual bullet-streams to strike Iraqi dictator Saddam Hussein's Presidential Palace, the headquarters of the Hussein-controlled Ba'ath Party, the missile complex at Taji, several vehicle depots, a military communications center, and a large radar array located in the suburbs, among other key targets.

Although most of these Tomahawks were armed with conventional, 500-pound, high-explosive "Bullpup" warheads, a few of them— including many fired from the decks of the USS *Wisconsin* on patrol in the Persian Gulf—carried a revolutionary new and classified electronics-warfare packet, dubbed "Kit 28."

Rather than decimate its target, Kit 28-armed Tomahawks made multiple passes over Iraqi power plants, spewing out thousands of tiny spools of carbon filaments. These filaments uncoiled as they drifted to the ground, to drape over power transmission lines and violently short them out. Loud pops and showers of sparks rained down near affected lines, to start dozens of small fires all around Baghdad.

Within a few minutes, entire electrical grids failed, and most radar sites and gunnery positions surrounding the Iraqi capital suddenly lost all their electrical power. In many places, the Iraqi-built, Soviet-designed anti-aircraft defensive system—thought to be one of the most complex, expensive, and formidable in the world—was completely overcome and effectively neutralized. From that point on, Baghdad was helpless and vulnerable in the face of the coalition attack.

THE MIGHTY TOMAHAWK

One of the many products and innovations of Ronald Reagan's arms race of the 1980s, a thousand Tomahawk cruise missiles had been built by McDonnell Douglas and delivered to the U.S. Navy by July 10, 1991. During the course of Desert Storm, 291 Tomahawks were fired at Iraqi targets. Most had a single high-explosive warhead, but 27 had cluster bomb-tips—large war-heads that burst open to spew hundreds of tiny bomblets over a wide area. All of these TLAMs were launched from the battleships *Missouri* and *Wisconsin,* and from several cruisers and destroy-ers. At least two submarines also fired Tomahawks into Iraq. An as-yet-undisclosed number of Tomahawks contained the effective Kit-28 warhead.

A post-war analysis of Tomahawk performance suggested that 85 percent of the 242 targets had been hit. Of those TLAMs that missed, two were shot down by the Iraqis, and the rest probably malfunctioned. During the Gulf War, the Tomahawk generally exceeded the U.S. Navy's faith in the weapon. The effectiveness of this weapon also astounded General H. Norman Schwarzkopf, who had openly expressed doubts about the Tomahawk's capabilities.

Despite the remarkable performance of the TLAMs over Baghdad, some key Iraqi radar sites were still active, and still posed a threat to the Coalition air attacks that were already well underway. Some of those sites, in a remote area of western Iraq, had been established to guard Iraqi Scud missile installations poised to attack Saudi Arabia and Israel.

The responsibility for smashing a safe corridor through this screen of protective radar, so that the Scuds could be bombed by the U.S. Air Force, fell to a flight of eight AH-64 Apache attack helicopters from the Army's 101st Airborne Division (Air Assault). These formidable gunships were led to their targets by two U.S. Air Force Pave Lows (helicopters with special avionics hard and software that allow them to perform in night operations or in bad weather) stuffed with sophisticated navigation and night-vision equipment, much of which is still nominally classified.

The AH-64 is the U.S. Army's principal rotary wing weapons platform, meant to replace the aging fleet of Vietnam-era AH-1 Cobra attack helicopters. Although the first Apache flew as far back as the late 1970s, more-recent versions feature many improvements over the earlier models, including the use of composite materials in its construction, and the addition of a state-of-the-art avionics systems.

COLD WAR SURVIVOR

Like so many weapons systems in America's arsenal, the Apache AH-64 attack helicopter was a Cold War design, intended to serve as flying artillery to stop the advance of Soviet tanks through the Fulda Gap in Europe in the opening hours of a World War III that never happened. The first Apache unit arrived in Europe in 1987 as part of President Ronald Reagan's revitalization of NATO's defensive capabilities. Since the fall of the Soviet Union, Bahrain, Egypt, Greece, Israel, Kuwait, Saudi Arabia, and the United Arab Emirates have all purchased and deployed the Apache.

At the core of the sophisticated electronics package inside the AH-64 is the Martin Marietta Target-Acquisition Designation Sight/Pilot's Night Vision Sensor (TADS/PNVS), which is linked electronically to the Honeywell Integrated Helmet and Display Sight System (IHADSS) worn by both members of the Apache crew. The sensor turret for the TADS and PNVS systems is located in the nose of the Apache and revolves to match the movement of the pilot's head. This turret also includes a Forward-Looking Infrared Radar (FLIR) sensor linked to the IHADSS. Target Acquisition imagery is relayed to a copilot/gunner console that has both a Head-Up and Head-Down Displays (HUD/HDD). The Head-Up and Head-Down Displays are flat plastic screens that provide critical information to the pilot immediately, so that his eye will not wander down to the cockpit controls and miss the action outside his canopy.

Gulf War Apaches contained three principal weapons systems: the McDonnell Douglas M230 30mm Chain Gun loaded with 380 rounds and fitted in a turret below the cockpit; Rockwell AGM-114 Hellfire laser-guided anti-tank missiles, which are suspended from the Apache's stubby "wings"; and two Hydra 2.75-inch, 70-mm rocket pods, also fitted to hard points on the Apache's wings.

Eight Apaches lifted off in the opening hours of the Gulf War, to fire the first shots of the conflict. Their targets were the early-warning radar sites in western Iraq. To accomplish their mission, the Apaches had to fly the nap of the earth—that is, only a few feet above the ground, rising only slightly to fly over hills, trees, or other obstacles—for a distance of 1,100 miles, over a frigid, night-shrouded, featureless desert waste, just to get within range of their targets. Two Air Force "Pave Low" helicopters served as pathfinders for this critical mission.

When the Pave Lows arrived within 12 miles of the primary Iraqi radar site, they dropped chemical flares near the target area. The plastic tubes made a highly visible greenish glow in the night-vision goggles of the Apache pilots, but were practically invisible to the naked eye. Mission accomplished, the Pave Lows retreated.

Following the path made by the chemical flares, the AH-64 Apaches surged forward to attack the Iraqis in the first engagement of the last large-scale war of the 20th century.

The Apaches achieved complete surprise, and within two minutes their gunners had scored 15 direct hits with Hellfire missiles. They had utterly obliterated the Iraqi radar installation, several mobile radar vans and support facilities, and cleared a safe-travel lane for follow-up forces to make precision strikes against the Scud installations and for 1,100 other Coalition aircraft to hit targets beyond.

Through the smoke and fire hovering over the ruins of the flattened radar installation, the Apache crews spotted the arrival of the Air Force strike package—two EF-111A electronic jammers, followed by two dozen F-15E Strike Eagle fighter/bombers and a flight of British Panavia Tornado Interdictor/Strike (IDS) aircraft. As the Apaches made a wide circle and headed back to Saudi Arabia, the Coalition aircraft bore down on the now-unprotected SCUD batteries. Using precision-guided missiles and dumb bombs, the warplanes annihilated the Iraqi missile bases in just a few short minutes.

Thirteen minutes after the Apaches began their attack on the radar sites, the first laser-guided GBU-27 bombs fell on Baghdad, dropped by U.S. warplanes. The GBU-27 is a steel-cased version of the standard, 2,000-pound Paveway III Low-Level Laser Guided Bomb (LGB). The difference was that these particular GBU-27s were specially configured to fit inside the bay of an F-117A Nighthawk, the first active-duty stealth fighter in the world.

Smart bombs like the Paveway series, which can be guided to their targets by a laser-targeting system, are not a recent development. In fact, the first generation of Paveways entered service in the late 1960s, during the Vietnam War. Over 21,000 of them were dropped on North Vietnamese targets in the final two years of that conflict—more than were dropped during all of the Gulf War. What was new in the Persian Gulf War was that the most advanced generation of "smart bombs" were delivered to their targets by F-117A Stealth Fighters—the first-ever marriage of *stealth* and *smarts* in a lethal union that would, as we shall see, ultimately revolutionize the battlefield.

On the night of January 17, 1991, the first GBU-27 ever dropped by a stealth fighter demolished half of the Iraqi Air Defense Center complex at Nukayb, 35 miles north of Baghdad. Less than a minute later, a second GBU dropped from another F-117A finished off the structure. During the Second World War, prior to the marriage of stealth aircraft and precision-guided munitions, demolishing a strategic target from the air with a mere two bombs was deemed impossible, and it was considered infeasible during the Vietnam conflict of the 1960s and 1970s. Suddenly, in the 1990s, such destruction was not only possible, but *likely*, as nearly 90 percent of precision-guided bombs hit their intended targets.

In all, Nighthawks delivered 31 percent of the precision-guided bombs that fell on Baghdad that first night. During all of the Gulf War, stealth aircraft flew 1,300 sorties, dropping more than 2,000 tons of bombs.

But on that first night, success was still uncertain. The supposedly radar-proof Nighthawks faced a crucial real-war test—they arrived over Baghdad in advance of the cruise missiles, so the Iraqi capital was still protected by active radar sites and an air defense system that included 60 surface-to-air missile batteries and thousands of anti-aircraft missiles. Despite that active radar and the presence of these advanced anti-aircraft systems, the Nighthawks flew into the guarded skies over Baghdad undetected, and destroyed their strategic targets, retreating from the scene before the Iraqis even knew what hit them.

The Lockheed Aeronautical Systems F-117A Nighthawk Stealth Fighter is a unique warplane based on a revolutionary concept: It is the first—but certainly not the last—product of a top-secret development program that began in the early 1980s at the Lockheed research facility nicknamed "the Skunk Works." An angular, single-seat, swallow-tailed aircraft, the Nighthawk is powered by two General Electric F404-GE-FiD2 non-after burning turbofan engines buried deep in the wing root to minimize noise. Nearly the entire aircraft is coated with a coal-black, radar-absorbing skin that has the texture of Styrofoam. The fuselage of the Nighthawk is also faceted with small, flat planes that are designed to deflect radar waves.

In order to minimize the aircraft's infrared heat signature, the engines are recessed and shielded by flat "platypus" exhaust vents that quickly mix hot gasses spewed from the engines with cold outside air. Rather than hanging from hard points below the belly or wings, the bomb payload is carried internally to further minimize the chances of radar detection.

Little has been revealed about the F-117s internal structure. Aluminum probably comprises the bulk of the craft's structure, although some elements may be made of a Dow Chemical fiber-and-resin mixture—perhaps the boron/polymer material called Fibaloy. The surface of the Nighthawk is said to be sheathed with tiles made of a classified Radar Absorbing Material (RAM), and the aircraft's nose and leading edges are rendered heat-absorbing and nonreflective by a still-classified process. The tail section of the Nighthawk was originally constructed of metal on the first models to roll off the assembly lines, but these have since been replaced with a graphite thermoplastic surface after a near-fatal accident.

Avionics behind the glazed windows of the Nighthawk's advanced cockpit include the Texas Instruments Infrared Acquisition and Designation System (IRADS), which embodies FLIR (Forward-Looking

Infrared Radar), a laser designator, and Downward-Looking Infrared Radar (DLIR) systems. Both the FLIR and the DLIR are mounted in turrets, which are controlled by buttons on the pilot's joystick.

During an attack, the pilot of a Nighthawk closes on the target using the FLIR, then switches to the DLIR view in the final phase of the approach. The cockpit also holds a Kaiser Electronics Head-Up Display (HUD) adapted from the ones used on the F/A-18 Hornet, and a "four-dimensional" navigational system to ensure precision accuracy.

First deployed in Operation Just Cause to bomb the headquarters of the Panamanian Defense Forces 6th and 7th Infantry Division at Rio Tato during the U.S. invasion of Panama in 1989, the F-117A Nighthawk performed adequately using conventional "dumb bombs." But it was only after the Nighthawk was paired with precision-guided munitions on the first night of the Gulf War that this sophisticated weapons system realized its full potential.

The F-117A's primary target on that opening night of the Gulf War was Baghdad's International Telephone Exchange, a 12-story switching facility that handled more than half of the Iraqi military's telecommunications requirements and was nicknamed the "AT&T Building" by Coalition military planners. The target was deemed so important that two Nighthawks were devoted to the destruction of that single building. The Nighthawks performed perfectly, and the telecommunications complex was destroyed even before the first cruise missiles arrived over the capital.

Even as the telephone exchange was being leveled, another F-117A Nighthawk destroyed the 370-foot Al Khark communications tower. The explosion from the GBU-27 snapped the tall concrete structure in half, sending the upper portion of the steel structure crashing into the street. More stealth fighters attacked Command Bunker Number 2, Iraqi Air Force headquarters, several air defense command posts, the air-defense grid headquarters in Taji, and Saddam Hussein's country retreat at Abu Ghurayb. Of the 17 precision-guided bombs dropped in that first wave of attacks, 13 were direct hits. The rest struck near enough to their targets to cause significant damage.

As the night progressed, targets inside of Baghdad, all over Iraq, and inside occupied Kuwait were heavily bombarded by U.S. and Coalition warplanes. Thousands of bombs—both smart and dumb—were dropped from fighters, fighter bombers, and attack aircraft. Hard targets— buildings, bunkers, and telecommunications installations—were decimated by Paveway precision-guided missiles directed by aircrews circling high over the battle zone.

Individual Iraqi radar sites were detected and destroyed by swift, 1,000-pound AGM-88 HARM anti-radar missiles fired from U.S. Navy F-15 Tomcats. The HARM, with its digital autopilot and inertial navigation system, delivered a high-explosive punch that rode in on the tip of a low-smoke, solid-fuel rocket booster. The AGM-88 was an expensive addition to the U.S. Navy's arsenal, and was another result of Ronald Reagan's military buildup in the 1980s. Though the Navy often complained that the missile, manufactured by Texas Instruments, was far too costly to purchase and maintain, the AGM-88s more than proved their worth in Desert Storm by taking out more radar sites than any other weapon. Indeed, the HARMs radar-detection system was so sensitive, and worked on such a broad frequency spectrum, that it homed in on even passive radar emissions. The HARM gave Iraqi radar technicians two choices: turn off their radar systems completely, or watch them be destroyed.

Other targets were eliminated with the Navy's AGM-84E-1 Standoff Land-Attack Missile (SLAM). Derived from the highly effective Harpoon anti-ship missile, the SLAM is capable of two modes of attack: preplanned missions against high-value fixed or mobile land targets, and target-of-opportunity missions against ships at sea. In Desert Storm, the SLAMs were used against many different Iraqi targets, sometimes in pairs. The first SLAM would punch a hole in a target such as a large building. The second SLAM, arriving seconds later, would streak into that hole and destroy the target with an internal explosion.

Ninety minutes after the air strikes began, seven Air Force B-52G Stratofortresses launched a total of 36 AGM-86, Air-Launched Cruise Missiles (ALCMs) at predetermined targets deep inside Iraq. Like the Tomahawks, the AGM-86 guided themselves to their destination using advanced avionics and navigational systems. Though nuclear warhead-capable, the AGMs launched against Iraq were armed with conventional, high-explosive warheads.

The huge, 40-plus-year-old B-52s that delivered the ALCMs departed Barksdale Air Force Base in Louisiana on January 16, 1991, and flew for nearly 20 hours, to arrive at their pre-determined launch positions near the Iraqi border. Within a few hours of the initial attacks, the skies over Iraq were owned by the Coalition forces. Iraqi air defenses had been neutralized, and Saddam Hussein's Air Force had been swept from the skies. From that point on, Coalition fighters and bombers could range wherever they wished, seeking out targets of opportunity as they continued to degrade the once-formidable Iraqi military machine.

TOO BIG, TOO EXPENSIVE

When the first B-52 bomber was rolled out in April 1952, the prototype represented the most expensive investment in an aircraft by any nation since the final years of World War II. Echoing contemporary complaints about the B-2 Spirit stealth bomber, some critics pounced on the B-52 program, calling the long-range strategic bomber a financial boondoggle. Others wondered if such an aircraft would prove useful in the post-atomic age, especially with the coming of the first intercontinental ballistic missiles in the late 1960s. Despite these protestations, June 2005 will mark the B-52's 50th year of active-duty service. The B-52 has fought with distinction in the skies over Vietnam and in the Persian Gulf War. Today the aircraft is, on average, 25 years older than the crews who fly it, and the Air Force has provided constant hardware and software upgrades to keep the B-52 airframe current. After a half-century of active duty, the B-52 has proven its critics wrong.

All the while, sky-borne reconnaissance, sensor, and electronic warfare platforms operated high in the clouds, out of range of the blinded and crippled Iraqi air defenses. These eyes in the sky directed attacks and spied on enemy movements. No large force could move over land during the day or at night without being detected by these advanced Coalition reconnaissance platforms, or by satellites positioned in orbit over the Middle East.

The world was awed and amazed when the first images shot by precision-guided munitions, or the aircraft that fired them, were broadcast during a post-strike military briefing the morning after the Gulf War began. No newspaper, no magazine article, no Discovery Channel special had given Americans a clue as to how effective surgical strikes using precision-guided missiles and bombs would be. Pictures of missiles striking parked aircraft, racing between buildings to obliterate a military target, or flying through the open window of an Iraqi government building alerted the population of the world to the revolution in warfare brought about by precision-guided munitions. The future had arrived.

PART 2

GROUND WAR IN THE 21ST CENTURY

"The contest is always man to man, to end up with; everything in national defense is designed for that purpose and it has got to be that."

—William Mitchell in testimony before the House Appropriations Committee, 1921

CHAPTER 4

THE INFANTRY OF THE FUTURE

The foot soldier, whether he's called an infantryman, GI, or ordinary grunt, is the most basic element of combat, the smallest individual unit that collectively makes up the whole. Before there was a cavalry—whether horse, mechanized, or air—there were infantrymen who waged war with the distant ancestors of the weapons modern armies use today. Early foot soldiers fought hand to hand, with so-called shock weapons like the club, axe, or sword; or at a distance with projectile weapons like rocks, the throwing spear, the arrow, or the dart. All warfare revolved and continues to revolve around the needs and capabilities of the infantryman, for no victory is possible until the individual soldier can walk unopposed on to the field of battle. And so, a soldier's personal and physical needs have come to guide the evolution of warfare and its weapons. The fragility of the human body led to the development of the shield, the helmet, and body armor. The limits of human muscle and bone led to harnessing the horse, the elephant, the camel, and other animals to haul supplies and, in time, to carry the soldier himself into battle.

The rise of precision-guided munitions, air power, and highly mobile mechanized warfare seem to have diminished the importance of the foot soldier. Yet the extinction of the infantryman is nowhere in sight. In war, there are always places where airplanes cannot bomb, tanks cannot go, and precision-guided munitions are useless. Such places can only be fought over and captured by experienced foot soldiers using small arms. The task of the infantry is to seize terrain, and that job is likely to continue.

Urban environments, especially, are deadly for armored vehicles and impervious to air assault unless a large amount of collateral damage is acceptable to the attackers. When fighting in close quarters, in narrow, choked streets with an invisible enemy firing from cover, all the benefits that mechanized vehicles like tanks, infantry fighting vehicles, and armored personnel carriers provide are negated. There is no room for complex flanking maneuvers, nowhere for massed armored columns to move. Street fighting is the domain of the foot soldier, and will remain so.

THE BATTLE OF GROZNIY

The fatal vulnerability of armor in modern urban warfare is best illustrated by the Battle of Grozniy in Chechnya, where armored vehicles from the Russian Republic were shot to pieces by Muslim rebels while they tried to negotiate narrow city streets blocked with mounds of rubble. The problem for the Russians in Grozniy is a problem faced by all advanced armies in the 21st century—the lack of infantry strength in numbers required to take the city building by building in advance of the armor.

Urban warfare is grinding and costly, and any military faced with the prospect of house-to-house or building-to-building warfare better be ready to accept casualties. But it's less costly to send in foot soldiers to clear the way for armor than it is to send the armor in first. In the city, death for tanks and fighting vehicles is assured—from grenades, handheld anti-tanks weapons, hordes of charging soldiers or irregulars, or even Molotov cocktails hurled from high windows.

Since the American Civil War, the U.S. military has relied on technology to be a force multiplier. That is, a means to make a few men fight as if they were many, through the use of overwhelming firepower, such as heavy and light machine guns; handheld anti-tank and artillery pieces; and more recently, with precision-guided munitions. Unfortunately, since the end of the Vietnam War the U.S. infantryman has benefited from fewer high-tech, advanced solutions to the problems of fighting a war than his fellow seamen and airmen. This situation is about to change.

Today, the joint U.S. Army/Marine Corps/Joint Special Operations Command program known as Soldier as a System (SAAS) envisions a future infantryman who employs the characteristic stealth of his brethren seamen and airmen, combined with high mobility and the most advanced, deadliest weapons systems imaginable. This soldier of the future will blend in with his environment and be harder to kill than today's armored tanks. He will possess advanced multi-spectral sensors, high-speed computers that give him access to instant communications and battle management plans, and smart munitions similar to, but far superior to, those used by today's high-tech warriors of the air. With the deployment of the *Land-Warrior Infantry Integrated System* in the next 10 years, and the *Objective Force Warrior System* in 15 or 20 years, the U.S. military will take its first tentative steps in the evolution of the infantryman. Through the use of advanced technology, both the Land Warrior and the Objective Force Warrior will greatly overmatch his opponent, while greatly minimizing risk to himself.

The Land Warrior Infantry Integrated System looms just over the horizon. The Land Warrior will sport a high-performance Pentium computer worn like a belt around his waist that employs existing vision enhancement, direct communications, guidance, and weapons systems. This computer will be tied to a wireless local-area communications network and global-positioning and acquiring system, all accessible by a visor with a kind of super-advanced heads-up display (HUD) built into the helmet, similar to those used by aviators today. This ground war vision system is called a Helmet Mounted Image Display (HELMID), and it differs from an aviator's system in one fundamental way. Aviation HUDs are there to filter data to prevent information overload, which impairs a pilot's ability to make quick decisions in the blink of an eye— a requirement of super-fast air combat. The ground warrior's HELMID will be there to provide a vast array of battlefield data, more, perhaps, than he will ever need, but there in case he does. This is data overload to the *N*th degree!

Instant communications is the key to victory on the modern battle-field, and so the future soldier's wireless network will keep him in constant touch with his squad mates. By logging onto the network, the Land Warrior will always know the location of friendly troops, diminishing, if not completely eliminating, the chance of "friendly fire" casualties.

On his HELMID screen—located on the standard Kevlar "kraut" helmet just to the left of what looks like a pair of large, very dark sunglasses—the individual Land Warrior will be able to call up a map, superimpose his own position and the position of each member of his unit onto that map, paint a target with the laser designator attached to his rifle, or request intelligence data on the disposition of the enemy forces in the immediate vicinity. This information will immediately be projected onto the screen of his heads-up display, and may be disseminated by the commander on the field to the rest of his squad.

With instant intelligence available, the rapidly changing battlefield situation could be assessed every few minutes, and troops could be re-positioned to respond to emerging threats or to exploit an enemy's weakness. Such critical intelligence could originate from many sources: other members of his squad, a spy plane or drone hovering high overhead, satellite reconnaissance beamed from outside the earth's atmosphere, a man-portable ground-to-air radar system, or the thermal-imaging unit built into the scope on his rifle. With that same helmet, HELMID, and computer, the infantryman will be able to relay real-time data to headquarters, paint a target for an airborne attack, or call in suppressing fire by simply sending an e-mail. When calling for artillery support, the Land Warrior will be presented with options on his computer screen. Using one finger, he can designate the target location, type, and size—as well as the type of munitions desired, and the time on target—all in less than five seconds.

The thermal imaging system on the scope of the Land Warrior's rifle, which can detect the thermal (heat) energy emitted or reflected from an object from half a mile away, can also be used to "see" what the naked eye can not. Using thermal imaging, enemy soldiers are visible through darkness, through smoke or fog, behind trees, in a foxhole, or even through walls. This system is even more effective in detecting the heat radiated by heavy armor. With an advanced thermal imaging system slaved to his HELMID, the soldier need never expose himself to enemy fire. He only has to lift the scope of his rifle from cover, and then only for a moment. This system will be so effective that it also will allow a Land Warrior to detect an enemy soldier lurking around a corner by the thermal energy he emanates. The Land Warrior will be able to make the kill by using his scope and HELMID to target the enemy—again, without ever exposing himself to fire.

The U.S. Army and a team of scientific specialists from Oak Ridge National Laboratory in Oak Ridge, Tennessee, and other civilian and military think tanks across the country, predict that the more advanced *Objective Force Warrior* is less than two decades away. The goal of the Objective Force Warrior program is to develop a high-tech soldier with at least *20 times* the range and effective killing capacity of today's infantrymen. It is hoped that the first units of Objective Force Warriors will be deployed by 2015.

SCIENCE ON THE MARCH

The Oak Ridge National Laboratory (ORNL) in Oak Ridge, Tennessee, was established in 1943 to pioneer a method to produce plutonium for the Department of Defense. During the 1950s and 1960s, Oak Ridge built and operated several research reactors. In the 1970s, ORNL expanded to include research for the Department of Energy. Today, Oak Ridge employs more than 3,800 people, with approximately 3,000 more researchers spending two weeks or longer at the facility. With an annual research budget approaching $100 million, the Oak Ridge National Laboratory conducts research in several areas, including analytical and separation chemistry, fusion science and technology, instrumentation science and technology, radioactive beam technologies, nuclear physics and astrophysics, and nuclear science and technology.

According to George Fisher, National Security Director of the Oak Ridge National Laboratory, the Objective Force Warrior research is ongoing because "the Army wants to stretch the bounds of technology but still have something that is feasible and can be built."[1]

With the Objective Force Warrior program, the Army is seeking to integrate both emerging technologies and innovative combinations of existing technologies to improve the effectiveness and survivability of the infantryman of the future. In essence, the Army is seeking to create a technological "knight," armored, protected, and nurtured by the latest advances of modern science and technology. These warriors of the future will be capable of doing the work of 10, 15, or even 20 conventional troopers—a *very* effective force multiplier indeed. Such a force will have a vast area of operation. With sensor, intelligence, and combat support from unmanned aerial vehicles like the Global Hawk, a single squad— 5 to 8 men—could secure an area of up to 300 or 400 square miles. They could project explosive power over a radius of 50 miles with new smart handheld or robotic artillery, and call down unmanned air strikes on targets hundreds of miles distant from their own position.

The Oak Ridge National Laboratory—with its connections to the defense industry and military institutions and its knowledge of the latest in emerging technologies—has guided the "visioning process" of the Objective Force Warrior at the request of the Army. Concept-design teams composed of futurists, systems engineers, computer designers, biologists, military experts, speculative writers, medical personnel, and specialists in a dozen fields began work in 2001. In February 2002, they issued a report that featured a glimpse of the future Objective Force soldier.

Objective Force Warriors will be able to engage and destroy the enemy at longer ranges with greater precision and with devastating results. Technologies that would make such combat effectiveness possible include better communications than even the Land Warrior will possess, advanced situational-awareness software, built-in chemical and biological detection and protection systems, and advanced medical devices integrated into the "weave" of their all-weather, all-season battle uniforms.

Advanced weapons and targeting systems, including a helmet heads-up display (a further evolution of the HUD or HELMID) featuring more advanced software than exists today, will greatly enhance the Objective Force Warriors range of vision and killing power. Access to that computer may be through the use of hands, eyes, or even brain waves, where significant advances have been made. The U.S. Air Force has been working on a thought-controlled piloting system for more than a decade, and such breakthrough technology has amazing civilian and military applications.

Many advances will come with the deployment of a new generation of battle wear. Conventional Battle Dress Uniforms (BDUs) and the flak jackets of today will soon be replaced by an *integrated suit system* designed to protect a soldier and provide hemorrhage control in case a bullet penetrates the suit. Microfiber technology has already produced wearable materials that are light and that breathe, yet can seal themselves against toxins or inclement weather, cold, or heat. New fibers in development now can be used in any environment, and can change color to blend with their surroundings. Smart, "chameleon" technology has already been tested, and such materials may not only shroud the soldier of the future, but shroud his tanks, ships, and aircraft as well.

COLOR-COORDINATED BATTLE UNIFORMS

According to researchers at the U.S. Army Soldier System Center, in Natick, Massachusetts, the chameleon suit is closer than we think. By 2025, they predict, the United States will be manufacturing combat clothing that can change colors and patterns to mingle with their environment, and make fast temperature adjustments to blend in with weather conditions, as well. Best of all, these new uniforms will be breathable and 20 percent lighter than the standard BDUs of today.

Other innovations include clothing material that is flexible, but "reacts" to shock: If a bullet strikes or a bayonet stabs, the material instantly becomes hard as steel. Some of this technology is in development at NASA, but military programs are also experimenting with this technology.

Advances in medical technology have already given the military innovative new triage equipment. Today, for instance, the military employs inflatable triage suits, which are placed on soldiers after they are wounded. These suits expand with air to compress wounds and slow or halt bleeding, until extensive emergency medical care can be provided. The next generation of so-called "healing suits" would be worn by soldiers as they enter battle. Sensors built right into the fabric would detect the penetration of a bullet, shrapnel, or blade, assess the extent of injury, and provide instant triage by expanding to put pressure on a ruptured artery, for instance, or seal a deep gash so the soldier can keep on fighting. Portable biological sensors similar to the most advanced equipment used in the Fox Chemical Detection vehicle (see Chapter 8) in use today would, in the future, be miniaturized and integrated into clothing. These bio-sensors would be capable of diagnosing the extent of the injury and provide stabilizing medical treatment in the form of antidotes to biological or chemical toxins, or topical anti-infection or coagulation agents for simple wounds.

FASHIONABLE FIRST AID

Projected for the distant future are suits that sense what toxins they are exposed to and feed that information to a strap-on Pentium, as with the Objective Force Warrior, or to a remote computer that can formulate an antidote that the suit can, if necessary, manufacture and administer.

Connected to wireless communications networks, these same bio-medical sensors will monitor the condition of the troops on the battlefield, and the data will be transmitted to headquarters, where battle managers and commanders can instantly access information on the physical condition of his troops and what types of threats—whether chemical, biological, or conventional—they face. Sensors built into weapons will provide constant updates on ammunition supplies. Networking with field commanders through the HELMID and computer will provide data on the overall combat effectiveness and unit cohesion of troops in battle. Instantly gauging the amount of ammunition used in a given firefight, or the fuel or battery power expended, will be critical to the logistics of the modern battlefield. In the past, armies have been plagued by interruptions or mistakes in the chain of supply that gives ammunition-starved troops the wrong caliber shell, or provides hungry troops with new clothing rather than food. Such mistakes are less likely to occur as the military begins to catalog all of a unit's necessities by computer, and instantly dispatches replenishment squads to deliver the proper supplies at the right time. The instant availability of battlefield data will also save lives—infiltration, exfiltration, or evacuation of the wounded will be both quick and effective.

The helmet of the Objective Force Warrior will be similar to the Land Warrior's systems, but will probably be far more advanced and adaptable. The Objective Force Warrior's helmet will most likely be a sealed unit with a re-breather to clean and recycle the air, or with a toxic filter. It will contain communications, vision enhancements, and a laser designator for target ranging and painting, all accessible through the same heads-up display used by Land Warriors. The Objective Force Warrior's system will be different in that it will most likely be voice activated or eye activated and security keyed to the individual user. This "smart" helmet would enable a soldier to monitor power reserves, show the direction and range of an enemy contact, and provide an enormous amount of additional information, including the capabilities and deployment of the enemy in a radius up to hundreds of miles.

Using advanced Objective Force helmets, a single soldier may not have to call in artillery support, because he will control his own artillery, through instant communications with hundreds of robotic guns or self-targeting mines positioned in and around the battlefield. The Objective Force Warrior will also be able to guide and control the next generation of Global Hawk airborne drones. Using armed, unmanned aircraft circling on patrol overhead, the Objective Force Warrior can call down his own air strike, guide the aircraft in to its target, and decide what type of munitions to be used—all by sending instructions through his personal computer.

While many of the technologies to make the Objective Force Warrior a reality are available now, or are maturing, other so-called "breakthrough technologies" have yet to be developed. These include advanced fuel cells to power the computers, HELMIDs, battle suits, exoskeletons, lethal and nonlethal weapons, and lethal or nonlethal robotics. Also in the development stage is the critical nanotechnology that will make the tough, flexible, miracle fabrics of the future possible. The future of nanotechnology is particularly bright.

During World War II, the development of plastics helped the American war effort enormously. Every pound of plastic that could be used to replace strategic products like steel, aluminum, or rubber meant that those strategic materials could be used to help build another tank, airplane, or victory ship. Because of plastics, cockpits were reinforced with additional armor, which saved American lives. Parachutes made of nylon—another plastic—further saved the lives of countless airmen. In time, weapons like the M16 were made lighter and more rugged through the use of polymers.

In this century, nanotechnology will play a crucial role in the development of the new generation of army uniforms and equipment, and has the potential to transform future conflict in the same way that plastics changed the face of warfare in the last century. Nanotechnology is the science of manipulating particles smaller than 100 nanometers—or $\frac{1}{1,000}$ the width of a strand of human hair—to create new materials with sometimes amazing properties. By introducing tiny nanoparticle reinforcements into polymers, for instance, helmets and tank armor can be made 40 to 60 percent lighter. Nanotechnology will be used to create the "smart" uniforms for Objective Force Warriors.

Although the breakthroughs researchers hope to make with nanotechnology sound like science fiction, Tom Tassinari, a scientist with the Soldier Systems Center, is optimistic. "We are in the early stages of anticipating how nanotechnology will revolutionize army equipment .... Research in the field is already showing tremendous promise."[2] And according to Dr. Mike Sennett, also at the Soldier System Center, "With nanotechnology, we can add properties to materials that weren't there before ... (with it) we aim to help soldiers do everything they need to do with a smaller amount of equipment and a lighter load."

One look at the military research funding makes it obvious that nanotechnology is one of the military's key areas of focus, alongside chemical and biological agent detection and high-energy lasers. In addition to awarding $8.75 million in research awards to academic institutions conducting nanotechnology research, the Department of Defense has announced its intention to create a new research center called the Institute for Soldier Nanotechnology. This institute, which will develop within an existing university, is expected to receive about $50 million in funding in the first five years of its existence.

So what arms will the Objective Force Warrior carry into battle? Military planners believe that the small arms of the 21st century won't be so small after all—in fact, the foot soldier of the very near future might carry more firepower than an M1A1 Abrams—the Army's Main Battle Tank—has today. Many predict that man-portable artillery, smart munitions, exoskeletons, battlefield robots, and advanced battle suits will render low-yield, line-of-sight projectile weapons like the rifle, handgun, and machine gun obsolete. The systems that guide most precision-guided munitions to their targets are getting smaller each year. With advanced miniaturization, it will soon be more desirable for an Objective Force Warrior to carry a smart bazooka, smart rockets, and a launcher, or even a man-portable cruise missile, than to carry conventional small arms with "dumb" bullets that rely on aiming and line-of-sight flight to be effective.

A vast array of new weapons will be available for the Objective Force Warrior to bring into battle—that's *bring*, not carry, because most of these weapons need not be carried at all. Instead, they will be deployed and defended by battlefield robots positioned on the battlefield ahead of time or alongside the future warriors as they move into the area. These robots and their weapons may lurk many miles away from the soldier himself, but that won't matter. Through instant communications and battlefield management, the Objective Force Warrior could fire the weapon from its remote location and guide it to the target by using the sensors in his helmet and the high-speed computer strapped to his waist.

Researchers also speculate about the future of "smart" bullets. Such projectiles would rest in a light, durable tube on the infantryman's back. Using his sensors, the soldier will scan his environment and feed the data into the tiny nanocomputers inside the projectile's "warhead." The bullet itself would probably resemble a miniature two-stage rocket. When fired, the first stage will blast the bullet from the tube—probably with a compressed CO_2 engine—and then be jettisoned. The warhead will proceed to the target, controlled by tiny nanotechnologically manufactured CO_2 thrusters, and guided by the nanochip sensor built into its tip. With internal guidance, such a bullet could go around a corner, over an obstacle, or through a small opening. The projectile itself will, of course, be explosive and probably timed to explode when it can do the most damage.

Along with man-portable and robotic artillery, the Army is working on advanced landmines, including the "Wham" system—an intelligent, robotic landmine that acts like artillery. This system can detect the approach of a target through vibrations in the ground, unearth itself, and aim and fire its anti-tank projectile or a brace of projectiles at enemy armored vehicles. New generation mines will even destroy themselves when the conflict is over.

A WELCOME DEVELOPMENT

Old-fashioned landmines, which lay in wait for a soldier or vehicle to run over them, will soon be a thing of the past—a positive development because buried conventional mines injure or kill an estimated 10,000 innocent people a year, according to UN statistics.

The soldier of the future will need to have Superman's vision to survive in the hostile environment of the coming battlefield. Extending the range of vision in space and in spectrum is the first hurdle to creating the future infantryman. Fortunately, a lot of multi-spectral vision enhancement technology required to make the HELMID work already exists, and more is on the way. Night-vision goggles, called IP (Image Intensification) devices, are standard issue to today's U.S. combat troops. They utilize ambient light, such as starlight, enhancing it thousands of times. These systems provide the vision to make warfare a 24/7 occupation. In the next few years, by gathering that same available light through a lens and converting it into electrons, and passing those electrons through a phosphor plate, that same ambient light can be multiplied *30,000 times*—making it possible for clear vision on moonless or overcast nights.

But future soldiers will have more than enhanced vision to help them. The National Research Council predicted in 1992 that future infantry sensors will include millimeter-wave synthetic aperture radars to provide high-resolution images that are responsive to the material properties of targets. Such systems can be configured so that the active and passive components share the same optics, and thus can provide pixel-registered images in a multidimensional space, which allows for multidimensional imagery. In addition, information and intelligence from other sensor platforms, such as satellites, unmanned aerial drones, manned reconnaissance aircraft, ground-based sensors, and other troops in the vicinity could be transmitted to the HELMID of the individual infantryman, extending his range of vision and field of operations exponentially. Man-portable artillery, armored battle suits, high-speed computers, personal arms, ammunition—how can the soldier of the future be expected to bring all this equipment to bear on the enemy? Robotics will certainly help. So will effective re-supply and support. But throughout history, infantrymen have been used as mules and, in the future, that is unlikely to change. There are limits to human endurance, however, and the Army is actively seeking ways to extend those limits.

An exoskeleton that would augment the strength of a soldier and enhance mobility, speed, endurance, range, and load-carrying capabilities has long been a dream of the Army, and the grunt in the field who too often feels like a pack mule. Author Robert Heinlein, in his classic 1959 science-fiction novel *Starship Troopers*, was the first to envision such technology. Now exoskeletal battle suits are closer to becoming reality than ever before.

The exoskeleton is a robotic-type device that can be strapped on or attached directly to the human body to add muscle power for heavy lifting or long-range running, walking, leaping, or even *flight*. Resembling the "powersuit" seen in the 1986 film *Aliens*, such human-assisting machines have been under development for a dozen years or more. The first exoskeleton was built at the University of California, Berkeley, a few years ago. The operator slips his hand through straps in two arm-like robotic grip devices, and the machine mimics and enhances the operator's movements through the use of hydraulic motors. It didn't take long for the military to recognize the value of such a device and begin research for how they could be developed for use in combat.

BATTLE-READY EXOSKELETONS

The Defense Advanced Research Projects Agency (DARPA) will spend more than $50 million over the next 5 years to develop an exoskeleton for combat use. According to DARPA, the exoskeletons will provide ...

- Increased payload capacity for individual soldiers, including increased firepower and ballistic protection, and the ability to carry more supplies, weapons, and ammunition into battle.

- Strength magnification with which to carry larger caliber weapons, clear obstacles, and increase the endurance of the individual soldier in the field.

- Increased speed and extended range, which will enhance battlefield reconnaissance and ensure dominance on the battlefield by providing larger areas of control through increased mobility and firepower.

In 2001, DARPA provided exoskeleton research grants to four facilities. Today, work on the development of the first military exoskeleton is ongoing at the Oak Ridge National Laboratory; the University of California, Berkeley; Sarcos Research in Salt Lake City, Utah; and Millennium Jet in Sunnyvale, California—which is specifically working on the development of an aerial version of the exoskeleton dubbed the *Solo-Trek Flying Platform*. DARPA calls for the first lower-body exoskeleton to be ready for trials in 2003, and a whole-body version ready by 2005.

According to a conversation with Hamayoon Kazerooni, professor of mechanical engineering and exoskeleton researcher at the University of California, Berkeley, the ideal exoskeleton is one that "the person will feel maybe 5 percent of the entire load of the backpack—the machine takes the other 95 percent."

Professor Kazerooni's Engineering Department is also working on a device dubbed the "Pogo-matic," which looks like a children's pogo stick kicked up a notch, with a compressed-air pump that gives the user an extra push against gravity. Using a descendant of this device, the future soldier could move at top speed by hopping.

The main obstacle to exoskeleton research is in the development of propulsion devices and the miniaturized power system needed to run them effectively in the field. Exoskeletons that run while connected to an electronic power source and an independent exoskeleton powered by a lawnmower motor have already been developed. But soldiers in the field can't take an electric generator with them, and a lawnmower motor lacks the necessary stealthy characteristics. More work needs to be done in the field of efficient, compact fuel cells before military applications can be found for exoskeletal technology.

Even with the best technology in the world, an infantryman's duration on the field is limited by the amount of food and potable water available to him.

In the near term, the water-purification techniques developed by the military over the last century will have to suffice. Systems like the Camelbak hydration system, the Storm, and the Thermobak for high-temperature/desert operations, have worked well in the field and have a long life ahead of them. But such primitive systems won't be enough to keep the Objective Force Warrior reliably hydrated, so the Army is working on a water filtration system that resembles an invention by popular science fiction author Frank Herbert, in his epic 1965 novel *Dune*.

A LESSON FROM *DUNE*

In *Dune*, the denizens of the desert planet Arrakis get much of their water through porous, skin-tight clothing called "stilsuits," which capture all of the body's waste moisture—including the dew of exhalation, sweat, and urine—and recycles it into potable water for drinking. Advances in nanotechnology will provide the Objective Force Warrior with just such a "water suit" that will extend their duration in the field by many weeks or even months. Experiments to develop effective "hydration suits" are already underway.

With the problem of hydration dealt with, the focus will be on food and nutrition. Just in time for the Objective Force Warrior's debut will come the replacement for the MRE—the military's pre-packaged *Meals, Ready to Eat*. Already MREs have been developed that specifically provide the nutrition and caloric intake necessary for optimum performance in targeted climatic conditions; the MCW (Meal Cold Weather)/Long Range Patrol family of rations, for instance, is essentially the same high-quality freeze-dried food available commercially for campers. But the military version of such rations is immune to freezing, and includes a higher caloric diet and specific vitamins and minerals for performance enhancement in cold-weather conditions.

Performance-Enhancing Ration Components (PERCs) have also been available for more than a decade in the form of sports nutrition bars and sports drinks. Today's Army also employs performance-enhancing rations, but they are very different than their civilian counterparts. Army PERCs are not only formulated to provide energy and forestall fatigue, but they also include brain stimulating and attention-enhancing experimental drugs to improve situational awareness and other mental faculties during times of high stress—and there is no higher stress than physical combat.

Right now the Army provides ERGO drinks that provide replacement minerals, vitamins, and other crucial nutrients. Packed in pouches that, when mixed with water, make a quart at a time, ERGO is popular with troops and is heavily used. Also available is the Hooah Bar—a candy bar-size ration packaged like a commercial nutrition bar that can be consumed on the run—and MERCs (Mobility-Enhancing Ration Components), a family of food products that look like a large candy bar, or a pastry filled with meat, cheese, or vegetables. MERCs provide enough nutrients to help keep a soldier's belly filled and satisfied during the mission.

But when a soldier is deployed to the field for as long as the Objective Force Warrior may be, even these nutrition systems could be exhausted. That is why the Army is busy trying to develop the next generation of feeding product—the *nutrition patch*.

Resembling a nicotine patch used to help smokers stop their habit, nutrition patches placed on the soldier's skin will excrete a balance of vitamins, minerals, and nutrients required to keep a soldier going for days or even a week, avoiding the necessity of eating—and voiding—altogether. Though not a satisfying way to receive sustenance, initial experiments with various types of food patches have proved promising. In the future, it is conceivable that soldiers won't complain about how bad the chow tastes—they'll most likely complain that their food patch itches!

If all these changes come about in the next 20, 30, or 50 years, the result will be the most dramatic transformation of warfare since the invention of gunpowder. The radical increase in firepower and accuracy available to the Objective Force Warrior through man-portable weapons and precision-guided munitions will make massed armies of hundreds of thousands of soldiers obsolete. For when dozens of weapons are fired in total accuracy by a few good men, armies will naturally shrink in size as they grow in quality.

It's quality vs. quantity on a human scale, for what we may be seeing is the end of the GI, the grunt, the common infantryman. The foot soldier becomes obsolete when masses of men are no longer required to fight wars. With nanotechnology, precision-guided munitions advances in communications, firepower, and robots, fewer and fewer troops will be required to wage war. In fact, the model for the soldier of the future is not the citizen-soldier of America's past, but a highly trained, experienced, professional, technological warrior—*the Special Operations soldier*.

CHAPTER 5

THE ELITE WARRIOR:
ADVANCED SPECIAL OPS FORCES

Kings, queens, princes, and warlords have always singled out
an army unit, regiment, or guard for special favor. The Caesars
had the Praetorian Guard, King Henry his archers, Winston
Churchill the British Commandos, and John F. Kennedy his
"Green Berets." Today, America's elite, special operations forces
have moved to the front lines in the War Against Terror—they
were the first U.S. troops deployed to Afghanistan to dislodge
the Taliban and to hunt members of al Qaeda. Yet few Americans
know what it is that makes the special forces so special.

By definition, special operations forces possess talents and
capabilities that far exceed the abilities of the average front-line
infantryman. An elite corps of noncommissioned (NCOs) and
commissioned officers, special operations forces receive special
physical and mental training. A constant, rigorous screening
process whittles away at recruits until only the best and bright-
est remain. Special forces units—called A-Teams, Alpha Teams,
or Direct Action Teams—are not organized, equipped, or trained
to conduct sustained combat operations alongside massed armies.
Their units are small—from 5 to 15 men—and their missions
are of short duration, with limited goals. Special forces don't
fight on the front lines. Most are trained to operate deep
within enemy territory. Intimately familiar with the culture and

terrain in which they operate, some members of the special forces can pass as natives within their areas of operation.

Special ops forces don't rely on overwhelming force to defeat their adversary. Some special operations specialists never see combat at all. Psychological operations units, for instance, wage war from a distance, through white—overt—and black—covert—propaganda, electronic interference, fear, intimidation, and misinformation. Direct action teams use stealth, speed, and the element of surprise to achieve their objectives. Others train indigenous peoples to battle America's enemies. Because their numbers are few, special operations forces do not annihilate the enemy, but *undermine* them.

Elite special operations forces are expensive to recruit, train, build, and maintain. Their ranks are small, because the pool of talented, patriotic, intelligent, self-motivated individuals who meet the rigorous physical, psychological, and mental standards of the special forces is limited. Despite their size, special operations units receive a dispropor-tionately large slice of the defense budget. This lavishing of resources is consistent with the Pentagon's ongoing preference for quality over quantity in men and machinery.

The first modern special operations forces appeared in World War I. German light infantry shock troops, or *storm troopers*, were trained to penetrate the Allied lines and interfere with command and control on the eve of the spring offensive of 1918. In World War II, the British, Germans, Canadians, and the U.S. Army, U.S. Army Air Corps, and U.S. Navy developed and maintained special warfare units to perform various, specialized tasks that would be defined today as "covert or spe-cial operations."

The unwanted stepchild of the conventional military, special forces units from previous wars have been used and abused, only to be dis-banded at the close of hostilities. That situation changed somewhat during the Cold War. Facing nuclear Armageddon—and a series of proxy wars against the Soviet Union in the form of communist insurgencies in Cuba, Iran, Korea, and Vietnam—the U.S. military searched for a new way to counter the threat of communist revolution. One strategy that was developed during the Eisenhower administration involved the establishment of the First Army Special Forces Group at Fort Bragg,

North Carolina, on June 20, 1952. Members of this premier unit included young recruits from Eastern European nations who spoke their native tongue and were familiar with the socialist system and Soviet tactics.

Enamored by new methods of unconventional warfare developed by the European colonial powers, and impressed with Britain's successful use of counterinsurgency tactics to suppress a Communist uprising in Malaysia, President John F. Kennedy elevated the status and importance of the Army Special Forces in the early 1960s. Soon, these elite soldiers became known for their distinctive headgear, and the legend of the Green Berets was born. JFK dispatched the Army Special Forces to Vietnam, where their exploits were celebrated in story, song, and on the screen. Suddenly, the Green Berets were the most famous soldiers in the world. Funding increased along with public visibility, and other armed services followed the Army's example, creating special operations units of their own.

THE LEGEND OF THE GREEN BERET

When President John F. Kennedy visited the Special Warfare School at Fort Bragg in October 1961, he requested that all special forces troops present at the event wear their still unofficial and unauthorized green berets. One year later, during a second visit to what would become the John F. Kennedy School of Special Warfare, President Kennedy called the Army Special Forces' distinctive headgear "a symbol of excellence, a badge of courage, a mark of distinction in the fight for freedom."

So it was that the legend of the Green Berets was born. In a few short years the Army's previously neglected special forces unit became the most famous military outfit in the world. Though untested at the time JFK lauded them, the Green Berets proved their mettle in Vietnam, at places like Nam Dong and Son Tay.

During the 14-year war in Southeast Asia, Special Forces troops would earn 17 Medals of Honor, 60 Distinguished Services Crosses, 814 Silver Stars, 13,234 Bronze Stars, 235 Legions of Merit, 46 Distinguished Flying Crosses, and literally thousands of Purple Hearts.

In 1961, the U.S. Navy, by presidential mandate, revived its Underwater Demolition Teams (UDTs). The original UDTs were rashly disbanded after World War II, and their loss was felt during the Korean conflict. This new breed of underwater warriors was dubbed the Navy SEALs (Sea, Air, Land Teams) and forged a legend of their own. In 1977, the Army's response to international terrorism—Delta Force—joined the ranks of America's elite fighting forces. And in 1990, the Air Force revived its Air Commando program in the guise of the Air Force Special Operations Command. Only the Marine Corps stayed out of this informal "arms" race, because, by definition, Marines are an advanced expeditionary force specializing in amphibious warfare. In other words, they are already "special."

The defeat in Vietnam was a particularly strong blow to the Army's and Navy's special operations forces. But the special forces muddled through the lean years of the 1970s without hitting the scrap heap. When the 1980s came, special operations did not benefit from the massive military expansion of that decade. Reagan's administration concentrated on weapons procurement and advanced technology—not on specialized troops or training.

Meanwhile, special operations forces participated in the military operations of the era with mixed results—from the failed hostage rescue mission in Iran in 1980 to the invasions of Grenada and Panama, and—despite General H. Norman Schwarzkopf's stubborn resistance—the Persian Gulf War of 1991. In Somalia in October 1993, special operations forces, including the 75th Ranger Regiment and Delta, served with distinction, and two Delta soldiers received posthumous Medals of Honor.

A major restructuring and unification of all special forces mandated by the Nunn-Cohen Amendment of 1987 further elevated special operations. As a result of this Congressional mandate, the unified United States Special Operations Command (USSOCOM) was created, and it now controls all special operations units of the Army, Navy, and Air Force, in essence establishing a de facto fifth branch of the U.S. armed services. With the creation of USSOCOM, Army Special Operations, Delta Force, Naval Special Warfare, Air Force Special Ops, the 75th Ranger Regiment, and the 160th Special Operations Air Regiment

(SOAR) all became members of a single, four-star, unified command with reliable and specifically designated funding.

The responsibilities of the USSOCOM are organizational—day-to-day military and security operations are broken down geographically into separate command, each of which handles special operations in specific regions of the globe. In this way, special ops teams can respond to a military crisis on land, sea, or in the air anywhere in the world, with just a few hours notice.

The Geographical Breakdown of USSOCOM is as follows:

- **Special Operations Command, South (SOCSOUTH)** is responsible for actions in Latin American and the Caribbean.
- **Special Operations Command, Atlantic (SOCACOM)** operates in the United States and the North Atlantic Basin.
- **Special Operations Command, Europe (SOCEUR)** works in Europe, West and Southern Africa, and Israel.
- **Special Operations Command, Central (SOCCENT)** is responsible for actions in Central and Southwest Asia, East Africa, and the Middle East (except Israel).
- **Special Operations Command, Pacific (SOCPAC)** performs in the Pacific Basin and Eastern Asia.

Under this new organizational structure, America's special forces retain their ties with traditional military branches—SEALs are recruited from the Navy, Special Forces from the Army, and so on. Each group also retains its original area of operation—Army special ops battles on land, the SEALs in the ocean and littoral regions, and Air Force Special Ops commands the skies. But much cross-training now occurs, and all groups must be airborne- (parachute-) capable.

In general, special operations forces are trained for specific types of missions, including …

- **Counter-proliferation (CP),** to prevent the spread of weapons of mass destruction.
- **Combating terrorism (CBT),** especially anti-terrorism and counterterrorism missions.
- **Hostage rescue** of American and foreign nationals held for political or economic reasons.

- **Foreign internal defense (FID),** especially training foreign troops to defend themselves against insurgencies.
- **Special reconnaissance (SR),** covert reconnaissance and surveillance activities.
- **Psychological operations (PSYOP),** dissemination of propaganda and misinformation.
- **Civil affairs (CA),** missions aimed at soothing the population of a nation where friendly military forces are going to operate.
- **Unconventional warfare (UW),** in which special ops forces actually wage a guerilla war against an enemy government.
- **Humanitarian de-mining,** helping underdeveloped nations in their effort to find and diffuse mines left over from previous conflicts.
- **Humanitarian assistance,** including famine relief or medical aid during epidemics.
- **Combat search and rescue (CSAR),** locating and exfiltrating friendly forces lost over enemy territory.
- **Counter-drug operations,** including interdiction and operations to curtail the production of narcotics.
- **Direct action (DA),** a fancy term for a traditional combat strike or raid to seize, capture, or destroy a designated target.

To accomplish these missions, special units use the most advanced small arms, man-portable artillery, night- and starlight-vision enhancers, high-speed computers, wireless communications, body armor, underwater breathing apparatus, helicopters, fixed-wing aircraft, surface and underwater boats, and parachute gear available. Today's special-forces units are required to become familiar with many different languages and cultures, and possess the skills to blend in to an exotic environment. They must also be familiar with the state-of-the-art systems of modern war, be physically fit, and be university graduates who possess rank and maturity. They must be computer literate and have good interpersonal and communication skills.

Many members of today's special operations forces possess advanced degrees—in medicine, psychology, sociology, computer sciences, oceanography, and a hundred other disciplines.

FOR MEN ONLY

Special forces soldiers are all men because, so far at least, the rigorous physical demands placed on America's elite units during training and deployment have prevented women from joining. This may change with advanced battlefield technology, but for the foreseeable future, special operations will remain the only all-male club in the military.

Once a recruit is accepted into the ranks of the special forces, he will expect to receive combat training far in advance of the other military branches—obstacle courses are tougher and demand more than physical skills to negotiate, situation and reaction exercises are more intense, team cooperation is stressed, specialized training is grueling, and constant physical and mental demands are placed on the individual until graduation and beyond. Specialized "graduate" courses like Basic Underwater Demolition (BUD); High Altitude, Low Opening (HALO) and High Altitude, High Opening (HAHO) para-jumping; and Survival, Evasion, Resistance, and Escape (SERE) training are conducted under intense, combat-like conditions that further push special-forces troops to their limits.

In the near future, *urban-warfare specialists* may be added to the roster of skills and mission profiles of America's special forces. In fact, urban warfare will most certainly be a growth area in the military in the coming decades. Many factors point to this trend. Urban warfare is manpower intensive, and military strategies must be adopted that reflect the many dimensions of city fighting. Mechanized warfare on open ground tends to be horizontal and mobile. Urban warfare is vertical, because movement is not only back and forth, but *up* into the upper floors of buildings and skyscrapers, or down into cellars, sewers, and bunkers.

In the future, specialization in the field of urban combat will become a requirement for special ops warriors, and new weapons—including personal armor, specialized types of armored vehicles for urban fighting, and nonlethal weaponry—as well as new tactics will be needed for the armies of the future to clear an urban environment of enemy forces.

More specialization not yet envisioned will also likely occur as warfare continues to evolve. The end result of all this training, all this expertise, is a true new-model warrior—a man familiar with the traditional strategy of tactics of asymmetrical warfare, able to speak foreign languages and feel comfortable within exotic and often primitive cultures, but who also possesses the skills and expertise to understand and operate the most advanced small-arms, computer, communications, and man-portable munitions on the cutting-edge of technology.

Today's special-forces soldier represents the long-dreamed-of marriage of the technician-scientist (as embodied in World War II by the British "boffin" and in the Cold War by the Rand Corporation analyst) with the skilled, highly proficient professional soldier. Such a fusion was unimaginable in the 1940s and 1950s, but is a requirement in modern warfare. Tomorrow's special-operations soldier may resemble an armored superhero with amazing powers of destruction at his fingertips.

So the future of America's elite special forces is bright. Because they are unified under a single command, and because the missions for which they were trained and the elite skills they possess so closely reflect the strategic needs of the U.S. military in the 21st century, it will be much easier for future chief executives to turn to special forces in times of crisis. That is exactly what George W. Bush did after the attack on the World Trade Center in 2001. This trend is likely to continue, as skirmishes between West-hating terrorists and civilized societies increase.

This shifting trend may mean that the next revolution in the future of warfare will be a *human* revolution as much as a technological transformation. Within the next 20 years, special-forces soldiers trained to the highest levels of combat proficiency and technological expertise will be united with high-speed computers and instant communications; with their own strike-capable drone aircraft; with advanced, precision-guided, man-portable artillery; with robotic weaponry and aids; with nanotechnologically enhanced clothing and equipment. The result may well be an invulnerable super-soldier who will transform the face of future warfare.

THE RETURN OF THE WARRIOR ELITE?

While many people herald the evolution of the super-soldier, others fear just such a revolution, and worry that such specialized training combined with exotic, complex, high-tech weapons will lead to the rise of a new class of warrior elite, resembling the knights of old, or the samurai class of feudal Japan. The long-term existence of this professional warrior class may reintroduce the concept of a warrior elite. By design, mass armies are democratic armies. But small units consisting of skilled and brave professionally trained warriors with enormous combat power at their fingertips represent a challenge and a threat to America's democratic ideals. As George and Meredith Friedman warned, "Meritocracy may well turn into aristocracy."[1]

Such developments might bring about another social shift. The shrinking of the size of armies without diminishing their military might well contribute to the re-emergence of the powerful city-state or small nation-state.

Future conflicts and precision-guided weapons won't require masses of soldiers to use them; instead, they'll require ability and expertise. With the emergence of formidable new technologies of war, small, wealthy nations with an educated pool of recruits and a good industrial base could become the future's military superpowers. Nations like Singapore, Taiwan, or Israel can produce the instruments of war and the professional soldiers required to use them, and may emerge as major military powers in the future of war.

For the first time in five centuries, Western civilization is about to witness a dramatic and permanent decrease in the size of its land forces, without any decrease in military power, as mass armies of citizen-soldiers are slowly and inexorably replaced by an elite cadre of military forces that are truly *special*.

CHAPTER 6

CUTTING-EDGE VEHICLES, ARTILLERY, AND WEAPONS

Soldiers have always sought mobility in battle and as a means to crash through enemy lines. The mounted cavalry fulfilled that important role from the 8th century B.C.E. until the middle of the 20th century. At first, horses were used to pull chariots; after horses were bred large enough to carry a man, the first cavalry soldier made his appearance. In time, and with innovations like the stirrup, the cavalry—especially armored or *heavy* cavalry—came to dominate warfare. In medieval Europe, the mounted warrior became the most formidable military instrument of the period.

In World War I, European cavalry emphasized surprise and mobility. But mounted troops had little chance to break an enemy armed with machine guns and heavy artillery. The horse-cavalry's failure to produce a breakthrough on the Western Front would mean its extinction. Although the use of mounted cavalry persisted through World War II—most notably on the Eastern Front where such troops could occasionally execute a successful shock action—horses were replaced by a new mode of transportation: the internal combustion engine.

The first mechanized vehicle used in combat was the armored tank.

Initially designed as an armored bulldozer, the tank swiftly evolved into an integrated weapons system designed for breakthrough, exploitation, and pursuit. Although regarded as the epitome of land warfare in the 20th century, tanks have only ever been used effectively as part of a combined-arms team. If tanks are deployed without air support or infantry, they are vulnerable to attacks from the air, from opposing infantry armed with anti-tank weapons, or from other tanks.

THE INSPIRATION FOR THE TANK

The word *tank* was coined in an attempt to disguise the true purpose of the first armored "land leviathans." The concept of the tank dates back to Leonardo da Vinci and in modern times, H. G. Wells. Before World War I, two military thinkers, Austria's Gunther Burstyn and Australia's Lancelot de Mole, produced designs that were tanks in all but name. But it wasn't until the stalemate on the Western Front in 1917, as the British sought innovative new ways to break through the German trench line and restore mobility to the stagnant conflict, that the first tank was developed. The Mark V was designed to crash through barbed wire and bridge trenches and to carry machine guns and cannon close to the enemy. The Mark V came in both a so-called male version (with cannon) and a female version (machine guns only).

Among the many inventors and strategists who had a hand in the tank's invention were British future-war-fictioneer Ernst Dunlop Swinton, and J.F.C. Fuller, once a disciple of British occultist Aleister Crowley who became one of the foremost proponents of armored warfare between the two world wars.

During World War II, the Germans most successfully deployed the tank in their blitzkrieg conquest of Poland, Holland, France, and Belgium. Perhaps the German army's greatest innovation came with the union of the tank and the radio, which allowed for effective battlefield management and rapid intelligence gathering. In 1940, the tank was still regarded as an anti-infantry, anti-fortification, highly mobile weapon. But as the war progressed, the tank evolved to fit an antitank role. At Kursk and in the two Battles of the Ardennes, massed armies of tanks squared off in mechanized duels.

Meanwhile, a specialized arms race was taking place, as each side built tanks with more efficient armor and larger guns. Yet, the status quo that existed at the beginning of the war never changed: The Germans built a small number of precision tanks that could outshoot and out-maneuver Allied armor, while the Allies made so many tanks that they overwhelmed the Germans by sheer weight of numbers. But with each innovation, more resources had to be allocated to maintain the same levels of combat effectiveness. Armor got thicker, tanks got heavier, engines got larger, cannon became more powerful, but nothing really changed.

During the Cold War era, the Soviet Union built a massive tank force and concentrated it in Central Europe, where it was poised to sweep through Western Europe. The United States, along with the North Atlantic Treaty Organization (NATO) responded in kind, building tremendous tank forces of their own and positioning them in places like the Fulda Gap to counter the Soviet threat. Tanks were still designed along the lines of their World War II counterparts, with tracked drive (to distribute the weight more evenly than four wheels could); a revolving turret (to provide a 360-degree range of fire); thick and/or sloped armor (to repel bullets, anti-tank rounds, and artillery); and a large main gun. Crew size varied—three if an automatic loader was used, four or even five if not. A variety of new ammunition was employed, ranging from high explosive, to antitank rounds, to smoke bombs. The engines were usually gasoline or diesel powered. Considering the weight of the average tank, they could move quickly, attaining speeds of 40 miles per hour on smooth ground.

With updates to gun size and armor thickness, tank weight soon became a problem, and researchers have tried to lighten them in a number of ways. The British established the Directional Probability Variation (DPV), a massive study to determine where rounds were most likely to hit a tank. A mathematical formula was developed to determine probabilities. As a result, front and side armor was thickened, with less armor provided for the top and rear. This worked well for line-of-sight attacks, but made tanks vulnerable to air attack and infantry armed with anti-tank weapons.

Another way to diminish weight was to slope the armor, which served to deflect projectiles. In time, heavy steel and cast-iron armor was replaced with rolled homogeneous armor (RHA), which is tough and lighter than iron, but brittle and hard to shape. During the 1970s, ceramic armor produced from minerals such as boron carbide and silicon carbide was tested and found wanting. Like RHA, ceramic is brittle, which meant that it could withstand one or sometimes two strikes, but no more than that, at least not without shattering.

A breakthrough in tank protection came with the development of composite armor, consisting of layers of different materials—ceramics, steel, resins, polystyrene foam, and so on. Chobham armor, designed by British researchers and named after the facility that first produced it, is a type of composite armor that combines the advantages of steel and ceramic. America's premier main battle tank of the 21st century— the M1A1 and M1A2 Abrams—is sheathed in Chobham, among other materials, but the exact makeup of the various layers is a closely guarded secret. The true revolution in tank warfare arrived with the development of various anti-tank projectiles. From World War II to the present, tanks fired two types of rounds: kinetic and high-explosive. Kinetic rounds resemble bullets and rely on weight and velocity to provide the energy to pierce armor. As armor got thicker, the kinetic rounds got larger. To improve the killing effect of kinetic rounds even further, the armor-piercing, discarding-sabot round (*sabot* is French for "shoe") was invented. A saboted round spins on its axis as it leaves the rifled gun barrel. The sabot falls away, leaving a small dartlike projectile that is stabilized by the spin. The dart slams into armor with its weight and velocity concentrated on a tiny point, allowing for penetration. Thicker or sloped armor can counter this innovation, so the armor-piercing, discarding sabot, fin-stabilized round was developed. This round can pierce thicker armor, but cannot be fired from a rifled barrel. This is because a rifled barrel has spiral groves cut into the barrel's interior, to give the projectile a spin as it is discharged. This technique is called "rifling" because it was originally developed for handheld rifles. These grooves, and the spin they give to the projectile, are what makes a rifle superior to a hand gun in both range and accuracy.

Another development was the high-explosive anti-tank (HEAT) round, which detonates prior to hitting the target instead of on impact. The premature chemical explosion is "shaped" by a copper-lined cone inside the projectile, which causes a jet of liquefied metal to extrude from the explosion. This extrusion behaves like a kinetic dart and pierces the armor like a blowtorch. A HEAT round can penetrate armor six times the diameter of the base of its cone. From the HEAT round there evolved the rocket-propelled, guided missile armed with a HEAT warhead. Add precision guidance to this mix, and you get a highly effective tank killing system that threatens the whole concept of armored warfare with extinction.

Other tanks, however, are not a tank's most ferocious enemy. In recent years, the helicopter has moved to replace ground armor as the "queen of the battlefield." Man-portable anti-tanks weapons carried by infantry troops are also a threat to the tank's survival on today's field of battle. In the 1970s, Israeli tanks were decimated by new Soviet-made, hand-held Sagger anti-tank weapons that were far superior to the bazookas of World War II.

Tube-launched, optically tracked, wire-guided (TOW) missiles were America's answer to the Soviet AT-3 Sagger—and both were descendants of the German X-7 *Rotkappchen* of World War II. These hand-held or tripod-mounted weapons feature projectiles connected to a control system by a long, thin wire. TOWs and Saggers don't depend on simple ballistics for accurate trajectory, because the wire provides in-flight control, which means that the shooter can guide the projectile to the target *after* it has been fired. The drawback of the TOW is that the operator must remain in the vicinity, and so is probably exposed to enemy fire, while he guides the projectile to the target. The benefit is that at 4,000 feet there is between an 85 and 96 percent probability of a hit—precision guidance, indeed.

More formidable is the next generation of TOWs, which will be deployed in the coming decade. The Line of Sight Antitank (LOSAT) system fires a shell that travels at speeds in excess of three miles a second, giving the target no time to maneuver. Precision-guided by a forward-looking infrared sensor located on the launcher, which transmits course corrections by laser rather than by wire, the LOSAT adjusts its trajectory through the use of small radial thrusters behind the nose. This revolutionary next-generation TOW does not need a tank to carry it—the LOSAT can be fired from small vehicles, helicopters, or fixed-wing aircraft, and work is underway to equip the next generation of armored personnel carriers (APCs) like the Bradley Fighting Vehicle with LOSAT technology.

America's preeminent armored system at the dawn of the 21st century is the 70-ton M1A2 Abrams main battle tank (MBT). The M1A2 is a variant of the original M1A1, with updates to the interior control, navigation, and communications systems. This update has replaced the earlier M1A1 "Gulf War" variant. Both versions were manufactured by General Dynamics Land Systems Division in Lima, Ohio, the design was originally dubbed the MI Block II by the Tank Automotive Command (TACOM), which mandates the mobility and firepower specifications for all new military vehicle designs.

(Photos courtesy of General Dynamics Land Systems)

The M1A1 Abrams Main battle tank.

The Abrams' main armament is the 120mm M256 smoothbore gun, developed by Rheinmetall GmbH of Germany. The gun fires M865 TPCSDS-T and M831 TP-T training rounds, and M8300 HEAT-MP-T and M829 APFSDS-T high explosive combat rounds, which include a depleted uranium penetrator for added density and striking power. The main gun is also capable of firing a new generation of precision-guided tank ammunition (including smart, guided anti-aircraft shells) that is just now entering deployment. The first Abrams prototypes featured a main gun autoloader, but that system was eliminated because it was deemed slow, stupid, and prone to breakdown—a lesson about autoloaders the Soviets never absorbed. Textron Systems provides the Cadillac Gage gun turret drive stabilization system that keeps the gun trained on the target when the tank is moving—even at its top speed of 45 miles per hour and on rough terrain.

The three-man crew is led by the commander, who has a 12.7mm Browning M2 machine gun. The driver has his hands full controlling the metal monster, but the loader has an additional 7.62mm M240 machine gun. A coaxial 7.62mm M240 machine gun is also mounted on the right hand side of the main gun. For camouflage, a L8A1 six-barreled smoke grenade blaster is fitted on each side of the turret. A smoke screen can also be laid by an engine-operated system.

The M1A1 and M1A2 tanks incorporate steel-encased, depleted-uranium armor of the rolled homogeneous (RHA) type with armored bulkheads separating the crew compartment from the fuel tanks. The details are classified, but the outside armor is probably comprised of layers of RHA sandwiched with polyurethane foam that is backed with a layer of Chobham. Behind the Chobham there is most likely a layer of depleted-uranium mesh and another layer of RHA plate. The M1A2 advanced design features an additional layer of super-dense depleted-uranium armor sheathing much of the exterior. Hull armor ranges in thickness between two and five inches. The top panels of the tank are designed to blow outward in the event of penetration by a HEAT projectile. The Abrams is also protected against nuclear, biological, and chemical warfare.

THE ABRAMS GETS AN UPGRADE

In February 2001, retrofitting began to supply 240 M1A2 tanks with an advanced system-enhancement package (SEP) that contains an embedded version of the U.S. Army's Force XXI command and control architecture, new Raytheon Commander's Independent Thermal Viewer (CITV) with second-generation thermal imager, commander's display for digital color terrain maps, DRS Technologies second-generation GEN II TIS thermal-imaging gunner's sight with increased range, and the Driver's Integrated Display (DID) and thermal management system. This command and control system gives the interior of the M1A2 the feel of an advanced fighter aircraft. The U.S. Army has plans to deploy more than a thousand M1A2 SEP tanks by 2004.

The advanced, on-board digital fire-control system is the heart of the Abrams' killing power. The commander's station is equipped with six periscopes, providing a 360-degree view. The Raytheon Commander's Independent Thermal Viewer (CITV) provides the commander with independent stabilized day and night vision, automatic sector scanning, automatic target cueing of the gunner's sight, and back-up fire control. Naked-eye viewing is provided by three prismatic viewing blocks built into the top of the hatch.

The M1A2 Abrams tank has a two-axis Raytheon Gunner's Primary Sight-Line of Sight (GPS-LOS), which increases the first-round hit probability by providing faster target acquisition and improved gun pointing. The Thermal Imaging System (TIS) has magnification ×10 narrow field of view and ×3 wide field of view. The thermal image is displayed in the eyepiece of the gunner's sight together with the range measurement from a laser range-finder. The Northrop Grumman Laser Systems Eye-safe Laser Range-Finder (ELRF) has good range accuracy and target discrimination (real statistics classified). The gunner also has a Kollmorgen Model 939 auxiliary sight, which serves as a backup for the main system.

The fire-control computer automatically calculates the fire solution based on lead-angle measurement, the bend of the gun measured by the muzzle reference system, wind velocity measurement from a sensor on the roof of the turret, and data from a pendulum static-cant sensor—which judges the precise angle at which the tank rests on the ground—located at the center of the turret roof. The operator manually inputs additional information on ammunition type, external temperature, and barometric pressure before firing. Using this data, the fire-control computer gives the gunner the precise angle the cannon needs to be aimed in order to hit the target. The driver has two periscopes on either side and a central image intensifying periscope for night vision. The periscopes provide a 120-degree field of vision and include a DRS Technologies Driver's Vision Enhancer (DVE), and a Raytheon Driver's Thermal Viewer, which was first installed during the Gulf War. The driver operates the handlebar controls with the help of the DID panel.

Communications are enhanced by the Inter-Vehicular Information System (IVIS), a brand-new wireless network that connects each Abrams with one another and the distant mobile command center (MCC) where commanders and battle managers can accept data and direct the combat action.

For propulsion, both the M1A1 and M1A2 have a six-speed, 1,500 horsepower Honeywell AGT-1500 gas-turbine engine. The Allison X-1100-3B transmission provides four forward and two reverse gears. The M1A2 variant will eventually have the new LV100-5 gas turbine engine developed for the Crusader self-propelled howitzer system program, which was cancelled by the Army in 2002. This newer engine will be lighter and smaller and will feature rapid acceleration, quieter running, and no visible exhaust. The 490-gallon fuel tank gives the AG-1500 gas-turbine engines a range of 250 miles without refueling: The newer engine will be even more fuel-efficient, but exact specifications are classified. The Abrams is the first U.S. tank since World War II to be built completely from armor plate instead of a cast hull and turret. This method provides vastly superior level of protection and makes the Abrams easier to build, maintain, repair, and upgrade. Even the track is improved, with rubber pads that can be replaced without having to dismount the track itself.

In plain English, this means that the Abrams is perhaps the finest tank deployed today. But even with all its impressive systems, there is ongoing research to further increase the Abrams' ability to survive battle and stave off obsolescence. Indeed, obsolescence threatens the very concept of the tank, which is difficult to hide on the modern battlefield and presents a wonderful target for precision-guided munitions.

Advanced as it is, many insist that the Abrams is not the most advanced tank in the world. Instead, it is argued, that distinction goes to the Israeli Merkava. The Merkava is the only tank to have been created by tank soldiers, based on their own experience in Israel's conflicts. The key to the Merkava's success is crew protection. Major General Israel Tal, the Merkava's lead designer, reckons that 75 percent of the Merkava's weight participates in the business of protecting the crew. The screens and suspension system are made out of ballistic steel, and both the engine and transmission are placed in front of the tank with the crew stationed behind. A low entrance at the rear of the Merkava makes it possible for the crew to escape under fire without exposing themselves and to haul out wounded crew mates who may be unable to pull themselves through the upper turret hatch.

Ironically, further protection is provided by the fuel tanks. The crew is surrounded on three sides by the diesel fuel stored in a revolutionary "dynamic" tank. The impact of the projectile results in hydrostatical pressure, which turns the fuel itself into a resistant medium that pushes back at the projectile, reversing its energy in much the same way as reactive armor works. In fact, it was the Israelis who first pioneered reactive armor with the Blazer system of the 1970s. Reactive armor is made from panels of high explosives sandwiched between metal plates. These panels "react"—when struck, they explode outward, turning the projectile's energy against itself.

The Merkava's design is unique. It is the first modular tank—that is, passive armor is attached to the basic frame in modular form. Simply remove a few bolts and new armor can be added or damaged armor replaced. With a modular system, improvements in armor technology can be deployed immediately simply by applying the new armor to the original frame. This innovative design makes the Merkava one of the cheapest tanks in the world to build, maintain, and upgrade. There is no denying that the Abrams has an equal or superior fire-control system and gun, and superior munitions, too. But because the Merkava crew is so well protected, and the tank can survive many direct hits—by some accounts, up to six—the vehicle can move much closer to the enemy and thereby increase both its firepower and mobility. And because of the unique battlefield conditions in the Middle East, the Merkava was created to perform better in an urban environment than many other existing tank designs.

Improvements to America's main battle tank in the 21st century will no doubt incorporate advanced modular design, but will also include new types of smart-munitions and guns. Because chemical projectile weapons have reached the limit of their effectiveness, a revolutionary means of nonchemical propulsion will have to be found. The purest form of energy is electrical, which means that an electrothermal gun, or an even more powerful electromagnetic rail gun, may soon appear on the battlefield.

An electrothermal gun would generate a plasma jet inside a fluid by instantly superheating that fluid with a massive burst of electrical energy. The superheated plasma would expand, creating powerful waves in the fluid that could be used to force a projectile out of the gun barrel at a velocity much greater that what can be achieved with a chemical explosion. But how do we generate the amount of electricity required for such a gun, or deliver it to a thousand tanks deployed on the battlefield? Until an effective portable electrical generation system of prodigious capacity is invented, or until there is a breakthrough in battery and capacitor design that allows a tank to repeatedly produce electrical bursts of the magnitude needed to form a plasma jet, the electrothermal gun will remain an impossibility for use on tanks.

More promising is the electromagnetic rail gun, which uses a pulse of electrical current traveling down one rail inside the gun barrel, across an armature at the base of the projectile, and up another rail on the opposite side of the barrel. This creates a powerful magnetic field that accelerates the kinetic projectile to incredible speeds—up to 3,000 miles per hour. Although there are problems—a major obstacle is creating a kinetic round dense enough to remain intact at such velocities—the

U.S. Army has pushed for a working electromagnetic gun, and experiments continue on a 90mm electromagnetic prototype designed in mid-1990.

Meanwhile, the Army is about to debut its new generation of smart tank munitions, including the X-ROD, a kinetic energy round that contains a wave radar sensor and computer chip in its nose. Fired in the general direction of the enemy, the X-ROD scans the area for a possible target while in flight. After locating and identifying the target, the X-ROD fires a rocket motor, accelerating until it strikes the target.

The U.S. Army's artillery units are also using precision-guided special munitions, including the impressive new Jabberwocky round. The Jabberwocky "shell" contains a powerful broadband-radio noise jammer that disrupts enemy communications over a wide area. A deployed parachute slows its fall, and after landing it deploys a powerful antenna and starts working. The Army is currently developing a variation of the Jabberwocky round that can be fired by the gun on a main battle tank.

As tanks become faster and more powerful, the weapons to counter the threat of armor are improving as well. In the coming decades, tanks will need to become more mobile and more protected than before, because the next-generation artillery shells will prove a greater threat to tanks than ever. The Copperhead is the artillery's equivalent of a precision-guided bomb. When fired, guidance fins deploy and a laser seeker starts to scan for the target. Once located, the Copperhead projectile adjusts its course and homes in, striking with enough high explosive in its warhead to destroy a tank.

Another threat comes from Field Artillery Containerized Anti-Tank Mines (FASCAM), which are fired from artillery and deploy a number of smaller antitank mines capable of damaging a tank by throwing it off-tread, or destroying a wheeled vehicle. Fired from long distance onto an enemy-held crossroad or choke point, such aerial-deployed mines can disrupt enemy movement until they are cleared.

As FASCAM clearly shows, mine clearance will be critical on the battlefield of the future. The *Zeus System*, an armored Humvee fitted with a powerful laser beam designed to blast landmines from the battlefield, is currently being tested by the U.S. Army. The laser is mounted in a turret on top of the Humvee, and a soldier sitting at a console inside the cab manipulates a joystick to point a low-power green laser beam at the mine. The laser is similar to the types used for industrial cutting and welding, and delivers about 2,000 watts of heat, which detonates the explosive.

The challenge to building Zeus was preventing the laser from overheating. Industrial lasers are cooled by a continuous flow of water that wouldn't be available on the battlefield, so a closed-cycle cooling system has been developed. More challenging is the fact that a Zeus system effectiveness against deeply buried mines is limited, and is likely to remain so.

With the many types of sensors and precision-guided munitions (PGMs) that already exist or are about to be deployed, the future of the tank is in peril. Not even the much-cherished dream of U.S. Army planners—the all-but-impossible flying tank—will preserve the utility of such mechanized armor indefinitely. However, the biggest threat to armor's supremacy today and in the near future will not come from sensors or PGMs. It will come from the next "queen of the battlefield"—the low-level attack helicopter.

Today, the U.S. Army's AH-64 Apache is the finest attack helicopter in the world. With a seemingly ungainly shape resembling an angry wasp and the deadly sting of a scorpion, its armor and firepower make the Apache the equivalent of a tank, but one that can fly nape of the earth over the battlefield. The Apache is capable of day and night adverse-weather operations against enemy armor and hard targets like buildings, bunkers, and fortifications. It has the ability to sneak along at low levels, reconnoiter the battlefield, sort out targets, and launch weapons from a stand-off range outside the enemy's anti-aircraft cover.

Because the Army Aviation Center mandated that all new helicopter designs meet stringent standards of maneuverability, ballistic tolerance against enemy gunfire, and load-carrying abilities, the Apache entered service with a very high survivability rate—even on the ferocious battlefields of the 21st century. The tandem, two-man cockpit is armored to withstand a direct hit, and the windows are flat to lower the chance of glare. The AH-64 has crash-resistant, self-sealing gas tanks, is impervious to 7.62mm projectiles, can tolerate .50-caliber rounds, and can even survive against 23mm high-explosive projectiles so long as the engine, drive train, and flight control systems remain intact. If an Apache is brought down, it has an airframe structure designed to withstand an impact 20 times the force of gravity (20Gs) without killing its crew.

Apaches have an infrared signature suppression system called the "Black Hole" that thwarts enemy heat-seeking homing devices by mixing hot exhaust gasses with a large volume of cooler air, and insulates exhaust pipes so that the missile will not detect the hot metal. The AH-64 also contains a suite of electronic countermeasures including a radar-warning receiver to alert crews when they are tracked by enemy sensors, a radar jammer, chaff dispensers that release a cloud of metal strips to reflect particular radar frequencies and conceal real targets, flare dispensers to decoy enemy missiles, and an infrared jammer in the form of an electrically heated block on the tail boom that radiates so powerfully that the sensitive seeker head on an oncoming missile is confused and defeated.

The Apache's outer skin is mostly aluminum of semi-monocoque design, meaning that the skin and underlying ribs are formed into a single structure. It has two General Electric T-700-GE-701Cs engines capable of 1,800 horsepower each. With four overhead blades, the Apache is more efficient than the Vietnam-era, two-bladed configuration of the UH-1 Huey and AH-1 Cobra.

Most impressive are the Apache's vision sensors and attack characteristics. The Martin-Marietta Target-Acquisition Designation Sight and Pilot Night-Vision Sensor system (TADS/PNVS) is mounted in a rotating turret on the nose of the Apache, and its movement is slaved to the movement of the pilot's head. The crew's helmets are also highly advanced and enable the operator to aim the aircraft's weapons and sensors with a simple turn of the head.

At night, in fog or dust, or in adverse weather, the TADS/PNVS system is used in flight as well as for aiming the weapons. The pilot's view is displayed on a small round screen attached to the helmet in front of the right eye. This tiny screen also displays navigational and fire-control data. The two navigation systems include a NAVSTAR GPS receiver and the Litton Attitude-Heading Reference System (AHRS), which is standard in most Army choppers. Inertial reference works with an ASN-137 Doppler-Velocity Measuring System, a downward-looking radar that monitors the movement of the helicopter over the ground.

The AH-64 can destroy most battlefield targets with the M230 30mm chain gun mounted on its chin. The M789 shells fired are lightweight, and have a tiny shaped-charge warhead that can punch through several inches of armor. The rest of the Apache's weapons are hung from hard points beneath the stubby side-wings, and include Stinger missiles for air-to-air combat and a load of 2.75-inch/70mm rockets known as the Hydra-70 system. Hydra rockets can include a high-explosive warhead, smoke charge, illumination warheads, and even a flechette warhead packed with tiny anti-personnel projectiles resembling carpet nails. The stand-off and shoot weapon on an Apache is the AGM-114 Hellfire TOW missile, which can locate its target with an optical seeker in the missile's nose that homes in on laser light "painted" on the target by another helicopter, an infantryman on the ground, or—in the future— by a high-altitude drone reconnaissance aircraft.

Apaches are capable of firing multiple Hellfire missiles at many painted targets at the same time. Up to 16 tanks can be destroyed by a single, fully armed Apache from a range of 5 miles. And unlike the tank, the helicopter does not need line-of-sight to make a kill. The Apache can hide behind a hill, a copse of trees, a ridge, or a building, and can quickly unmask to launch its Hellfires.

The next generation of Apache, the AH-64D model, will be much improved due to an advanced Longbow system, a millimeter-wave radar designed to see both ground and air targets in any weather, day or night. The Longbow is mounted inside a disc-shaped array mounted on top of the Apache's main rotor. Because the Longbow is designed for stealth, it will be difficult for enemy sensors to intercept or detect its emissions. After a pop-up sweep of only a few seconds, the Longbow's target acquisition computer can detect, classify, and sort enemy vehicles into categories and project their position onto a digital map display. This data can also be transmitted to other units in the field or in the sky. Tied to the development of the Longbow is the Hellfire Longbow missile. Its classified systems will work with the Longbow radar to track and kill the enemy.

The Apache is clearly the battle tank of today's Army aerial fleet and will dominate the battlefield for the next decade. But the Apache is best used in conjunction with scout choppers like the OH-58D Kiowa Warrior Scout, which can forward deploy, locate, and paint the target for follow-up Apaches to destroy.

Right now, the Army is preparing for tomorrow's war and will soon deploy the next generation attack helicopter, the *RAH-66 Comanche*. First designated the Light Helicopter Experimental (LHX), the RAH-66 was meant to combine light attack characteristics with the properties of an advanced scout aircraft bundled together in a single airframe.

Stealth is the key to survivability on the postmodern battlefield, and the RAH-66 is the first bona-fide stealth helicopter ever built. Its all-composite fuselage—composed of fiber, carbon, and plastics—renders the RAH-66 nearly invisible to radar, with a signature smaller than a bird's. The Comanche is designed to pierce the fog of war through the use of advanced sensors, electronics, cutting-edge stealth technology, speed, and high maneuverability. Its task will be to find the enemy and destroy them without being seen.

The Comanche sports some unique features, including a ducted tail rotor called a FANTAIL designed by Boeing-Sikorsky, which serves to protect the vulnerable tail assembly, muffle sound, and mask propeller movement to foil enemy motion sensors. The Comanche will carry a two-man crew in tandem in a fully integrated, shared-control, sealed cockpit provided with an environmental filtration system for protection against nuclear, biological, and chemical attack or residual contamination. The chopper will be controlled via an array of programmable and rapid reconfigurable multifunctional displays, control stick, and pedals. The crew's helmets and sights will be slaved to many on-board systems, but will be more lightweight and advanced than those used by Apache crews, and may more resemble the HELMID system (see Chapter 4) with an added avionics package.

The RAH-66 power plant is provided by the advanced T-800 engine and manufactured by LHTEC of St. Louis, Missouri. The T-800 is very efficient, and variants and upgrades of this power plant will probably propel all of the Army's light and medium helicopters for the first half of this century. In the Comanche, the engine is buried in the hull to reduce noise and radar signature, and exhaust is vented downward through the tail boom, but only after it is mixed with cooler air to thwart thermal imaging.

When it is fully deployed in 2005, the Comanche will be the most heavily armed aircraft in history. The primary weapon will be a three-barreled 20mm Gatling gun with 500 rounds. Retractable door mounts in the side will lower to deploy internal weapons, including the AGM-114 Hellfire missiles, Hellfire Longbow missiles, Hydra-70 rockets, Stinger missiles, and several still-classified weapons that have not yet debuted. A pair of stubby wings similar to those on an Apache can be attached to the fuselage. These wings can mount drop-fuel tanks for an extended range of more than 1,200 miles, or additional weapons, though with these attachments the Comanche loses some of its stealthy characteristics. The impressive range will make the Comanche self-deployable to anywhere in the world, unlike the Apache, which must be folded up and packed into the belly of a transport plane to get to the battlefield.

In flight, the Comanche is highly maneuverable because it is the first U.S. military helicopter to make full use of a digital fly-by-wire control system similar to the F-16 Falcon fighter jet's—which means that all control surfaces are moved by small electric servo-motors instead of with an old-fashioned hydraulic system. The advanced FANTAIL rotor also allows the Comanche to turn faster than any other helicopter ever built. A competent pilot can maneuver the Comanche through narrow valleys or even through the concrete canyons of modern urban areas. Amazingly, the RAH-66 can also fly sideways at nearly 75 miles per hour. Top speed in forward flight is still classified, but probably clocks in at nearly 400 miles per hour.

The Comanche was designed with ease of maintenance in mind. An in-built fault-isolation assemblage informs the crew chiefs of exactly what is wrong with the complex internal systems and can isolate the problem for a quick fix. An engine change takes only an hour, and computer upgrades can be done with the addition of simple computer cards inserted into the electronics bay. In combat conditions, it takes only 15 minutes to service, rearm, and refuel an RAH-66.

With all these advancements, the RAH-66 Comanche is likely to become the premier front-line combat helicopter for the 21st century. Indeed, if all goes according to the Army's plans, with retrofitting and upgrades the Comanche will see at least five decades of service, rivaling the in-service duration of the Air Force B-52 bomber.

As of this writing, the Army has already deployed some advanced land-warfare systems that will become fixtures in the future of war. Some programs—most recently the Crusader Artillery Project—have been cancelled. But other high-tech weapons and vehicle systems have endured post–Cold War budget cuts and are now being fielded.

Armored personnel carriers (APCs)—tracked, ground-war vehicles to move troops rapidly to the battlefield—have been a part of modern warfare since the half-tracks of World War II. Today, the M2 Infantry and M3 Bradley Cavalry Fighting Vehicles are America's front-line troop-delivery medium. For the 21st century, the Bradley is being retrofitted with new command and control systems, and with the same Inter-Vehicular Information System (IVIS) inside the Abrams M1A1 and M1A2 tanks.

Another variant of the basic Bradley is the Bradley Stinger Air Defense Vehicle. An armored anti-aircraft system to defend forward-deployed armored forces, this APC will carry Stinger missiles and advanced radar instead of troops. But the long-term improvement to the basic Bradley design will be the M2A2 and M3A3 Bradley fighting vehicle, which will take existing M2 and M3 chassis, strip them down, and remanufacture a new vehicle on top of them. These vehicles will have advanced composite armor and some stealthy characteristics and may contain such innovations as an independent thermal viewer for the commander, and a laser range finder/designator for advanced reconnaissance and painting targets for attack, and a second-generation

Forward-Looking Infrared Radar (FLIR) system similar to the one used in the AH-66 Comanche. Retractable weapons that are buried inside the fuselage and deployed before firing may replace the top turret in this advanced APC, the details of which are classified.

Chemical and biological weapons will pose some of the biggest threats on the future battlefield. One of the most advanced forward-reconnaissance armored vehicles available to counter this threat is the Fuchs (Fox) Nuclear, Biological, Chemical Reconnaissance Vehicle, first deployed by the German army in the late 1980s and used by U.S. troops (under the designation M93) in the Gulf War. This vehicle so impressed American military planners that an advanced version of the Fox system has been permanently adopted by the U.S. Army and Marine Corps.

CHEMISTRY 101

Chemical warfare is the military use of lethal, harassing, or incapacitating chemicals designed to harm or kill. Chemical warfare began on April 22, 1915, with the German release of chlorine gas at the Battle of Ypres. After World War I, revulsion against the use of such weapons led to the Geneva protocol of 1925. During World War II, only the Italians and Japanese deployed chemical weapons, and sparingly.

During the Cold War, the Kennedy and Johnson administrations funded renewed production of chemical weapons, culminating in the use of defoliants and riot control agents in Vietnam, which resulted in international condemnation. In 1969, President Richard M. Nixon ordered the unilateral destruction of stockpiles of chemical agents, which led to the ratification of a 1972 treaty that forbade the further production of such weapons. In 1985, Ronald Reagan convinced Congress to fund chemical warfare research, but work focused on chemical protection.

By 1997, the United States joined the Chemical Weapons Convention, which not only banned chemical weapons, but also retaliation in kind. Nations that violate this ban will be punished by other means, most likely a nuclear strike. As it stands now, any nation that deploys nonlethal chemical agents is also in violation of this agreement.

(Photos courtesy of General Dynamics Land Systems)

The M93A1 Fox Nuclear Biological Chemical Reconnaissance System (NBCRS).

Powered by a 320-horsepower diesel V-8 engine, the Fox is a six-wheeled armored vehicle with a crew of four—a commander, driver, and two sensor technicians—who work in a sealed, filtered environment and do not have to wear chemical/biological protective suits. The Fox rolls on six oversized self-sealing or solid tires, a trend away from tracks first pioneered in South Africa with the fast, armored Merkat fighting vehicle. The Fox has a top speed of 65 miles per hour and a range of 500 miles, and is armor-protected against mortar, machine gun, and small arms fire.

In its rear compartment the Fox carries an integrated sensor suite, which can detect nuclear contamination. More exterior-mounted sensor systems detect and identify dozens of chemical warfare and nerve agents, from sarin to cyanide. Robotic arms gather samples from the ground with a silicone roller. An automatic mass spectrometer analyzes the results and, if toxins are present, the Fox plants a warning barrier of signs or transmitters to alert friendly forces of the danger, and more sensors to constantly monitor the environmental conditions. The analytical systems are directly connected to an inertial navigation computer that maintains and transmits an accurate and up-to-the-minute map of the toxic areas and their levels of contamination.

General Dynamics Land Systems, in partnership with the original builders of the Fuchs, Thyssen Henschel of Germany, are currently building the next generation Fox NBCs. The improved and advanced Fox/M92A1 is ready and will be deployed the next time the threat of nuclear, chemical, or biological weapons bare their lethal fangs.

A well-appointed, competently staffed mobile armored command post is vital to modern combat operations. The days when a commander could control the battlefield from the turret of a tank, with no more than a radio and code book in hand and a few runners to deliver orders, are long gone. Today, the XM4 Mobile Command Post has replaced the venerable M577 mobile command system that has been around since the Vietnam era.

The Army's spanking-new Command and Control Vehicle (CCV) has been designated the XM4 (also known as the CC4) and is built on the same chassis used for the almost-obsolete Multiple Launch Rocket Systems. But instead of launchers and rockets, the boxy XM4 CCV is stuffed with command and control systems, digital display flat screens better than you'll ever see at Best Buy, and lots of hardware and software that are still classified. The XM4 is powered by a 400-horsepower diesel engine and is protected by ballistic armor capable of stopping machine-gun fire. More important, the XM4 has an overpressure filtration system and air conditioning, and is shielded from nuclear contamination and from chemical and biological attack. Packed with computers, radios, radar systems, and remote control and battle management systems, the XM5 will fulfill all the needs of the busy battle commander on the go in the highly mobile and rapidly shifting battlefields of our savage times.

The XM5 Electronic Fighting Vehicle (EFV) is a variant of the basic XM design and uses the same chassis and boxy command center as the XM4. But the XM5 EFV is packed with the most advanced communications, radar, and electronic warfare systems, including jammers, and sports a rugged, quick-deployable antenna mast that's nearly 100 feet tall and a larger-than-normal power-generation system to feed all those hungry electronics. Inside the climate-controlled, environmentally protected crew compartment, six to eight technicians scan for enemy transmitters, follow artillery back to its source through thermal sensors, locate and target enemy positions, and distort or jam enemy communications and sensors. Of course, much of this is educated speculation—the actual range, power, and configuration of such systems are classified.

In the coming decades the U.S. military will require flexible, effective and efficient multi-mission forces capable of projecting overwhelming military power worldwide. To satisfy this requirement, the joint U.S. Army/Defense Advanced Research Projects Agency (DARPA) Future Combat Systems (FCS) program was developed to provide enhancements in land force lethality, protection, mobility, deployability, sustainability, and command and control capabilities.

The FCS develops combat systems that will be lethal, deployable, self-sustaining, and survivable in combat through the use of an ensemble of manned and unmanned ground and air platforms. These Future Combat Systems will be capable of adjusting to a changing set of missions, ranging from waging war to peacekeeping. An FCS-equipped force will be capable of providing autonomous robotic systems; precision direct and indirect fire; airborne and ground organic sensor platforms; precision, three-dimensional, air defense; nonlethal battlefield technologies; adverse-weather reconnaissance, surveillance, targeting and acquisition; and other critical combat functions.

Currently the FCS is concentrating on the continued development and improvement of unmanned aerial reconnaissance platforms, combat and military support robotics, de-mining technology, and improved technology for battlefield communications.

A vital component of the Pentagon's ambitious FCS program is exploring the use of battlefield robots. Boeing is the lead contractor in the FCS program, and is expected to subcontract some of their research to institutions like Carnegie Mellon University—a leader in robotics invention and design—and RedZone Robotics, a high-tech, cutting-edge technology firm in western Pennsylvania. According to a Defense Department statement issued in January 2002 in an effort to give the Army "a significant combat overmatch against all foreseeable enemies … extending through the 2025 timeframe" the work of the FCS program will concentrate on "a robotic, direct-fire system, a robotic non–line of sight system [as well as] an all-weather robotic sensor system … capable of adjusting to a changing set of missions, ranging from war-fighting to peacekeeping."

The resulting technology will "strike an optimum balance between critical performance factors, including … mobility, lethality, survivability, and sustainability." The Future Combat System program's quest for "robotic direct-fire system, a robotic non–line of sight system," and "an all-weather robotic sensor system" indicates that the move away from manned armored attack vehicles—whether in the form of planes, helicopters, or tanks—has already begun.

Already, the first advanced battlefield robotics systems have been deployed in the war against the Taliban in Afghanistan. According to an Associated Press wire story filed July 30, 2002, a robot called Hermes is the first to be tested. It has been exploring caves in search of hidden Taliban forces, stockpiles of weapons and ammunition, and critical intelligence that may have been left behind as al Qaeda forces fled the area.

Little Hermes and his cousins Professor, Thing, and Fester—the four prototype robots being tested—are each 1 foot tall and 3 feet long, and weigh in at close to 50 pounds. They are heavy enough to trip an anti-personnel mine, and tall enough to spring a trip-wire. The prototypes are tough enough to withstand the blast from such booby-traps, too. Each one costs about $40,000 and each comes equipped with a grenade launcher and a 12-gauge shotgun. The robots run on tan-colored treads that are rolled by bright-green solid-rubber wheels.

Hermes and his friends can negotiate rough and uneven terrain and deploy robotic arms to lift themselves over high obstacles. Between 2 and 12 cameras relay images to the robot operators—members of the 82nd Airborne Division—waiting outside the cave mouth. Operators guide these robots via a wireless joystick system and desktop computer. Hermes is capable of negotiating a mine field, and it can mark mines. The next-generation robot prototypes now being built will actually be capable of de-mining an area without endangering U.S. troops.

Hermes, Professor, Thing, and Fester represent the first-ever use of robots as tools of combat, and the program has solid backing— proponents believe that sending robots into caves, buildings, or other hazardous areas ahead of human troops will help prevent U.S. casualties in Afghanistan, and may be the key to a less manpower-intensive method of conducting urban warfare. And unlike in Vietnam, where soldiers called "tunnel rats" were sent into Viet Cong tunnels armed with only a pistol, for the new war in Afghanistan, robots replaced the human "rat," likely saving many lives. Hermes and his brethren are also being tested as a robotic sentries and listening posts.

These robots have also been tested in an advance reconnaissance role. The prototypes are fitted with a Global Positioning System and can see themselves and each other on a digital map, ensuring more efficient searches. The one drawback is range. The robots run on two six-pound rechargeable batteries that have a duration of only about an hour each. But at seven or eight miles an hour, one of these prototypes can cover just about the same amount of ground as a human scout—for two hours, anyway.

America's respect for life and the legacy of the citizen-soldier will guarantee that smaller armies will be deployed and minimal casualties will always be preferable to the mass-armies and mass-slaughter of traditional wars. So within the next 25 years, we will begin to see crews separated from their robotic tanks, robotic drones replacing manned attack planes, and mobile, robot-fired weapons replacing the armored tank.

Through the work of the Department of Defense and its Future Combat Systems program, we will undoubtedly see more smart machines capable of more and more sophisticated and complex combat actions. In a generation or two, military robots will seek out the enemy, identify a target, map the target's location, and initiate an attack.

And who knows what capabilities the generation of robots after these may possess? How far away are we from the Destroyer Droids of *Star Wars?* The police and security robots of *THX-1138?* The Daleks from *Dr. Who?* Or even the relentless robot armies of *Terminator* and *Terminator II?*

Add to this lethal, high-tech equation the battle-armored, leaping and flying troops foreshadowed in *Starship Troopers,* new ballistic weapons resembling futuristic ray guns, and the first- and second-generation of nonlethal technology entering deployment, and we get a picture of the array of weapons that will comprise the new world order of battle in the 21st century.

PART 3

FROM SEA TO SHINING SEA: NAVAL AND AMPHIBIOUS COMBAT

"A man-of-war is the best ambassador."
—Oliver Cromwell

CHAPTER 7

THE EVOLUTION OF SEA POWER

The gunship—from the galleon to the man-of-war, the ironclad to the battleship—was an invention born of necessity. The Atlantic seafaring nations, specifically England, Portugal, Holland, and Spain, were barred from the lucrative trading routes in the Mediterranean by the Venetian and Muslim monopolies. But these nations bordered on the Atlantic, and so developed formidable navigational skills and the requisite scientific knowledge and shipbuilding expertise to venture westward, into the Atlantic and eventually the Pacific Oceans.

After the European settlement of regions of North and South America, the Old World powers that operated these new colonial enterprises were forced to create navies of armed, long-range warships large enough to carry troops to protect their interests. The result was the gunship, a weapon system built to survive combat on the far frontiers of civilization, against a numerically superior enemy, in situations where there was no hope of reinforcement.

The English were the first to put cannons aboard vessels, creating the gunship. After that, the gunpowder-fired race to dominate the oceans began. Naturally, warships got larger and larger—and more expensive—with many more guns and crewmen added to each succeeding generation. This resulted in many innovations, like cannons mounted below deck in double rows to execute the broadside, but ultimately innovation reached absurdity with cumbersome monstrosities like the Portuguese galleon the *Sao Joao* and its 366 guns. Faster and sleeker proved better when a mighty Spanish Armada of ungainly but well-armed warships was defeated by an out-gunned, but more maneuverable fleet of British vessels. The rest, as they say, is history—the history of British ascendancy.

The gunship continued to dominate sea power for 500 years, with ships evolving along with technology. Because of their far-flung empire, the British were especially keen to add the latest innovations to their armada. They were the first to add coal-burning steam engines, and the first to clad their ships in iron (see Chapter 2). In the 19th century, the gunship was replaced by the battleship. Its speed, armor, and fire-power made it the perfect platform to project political and military power across the oceans, to the remotest regions of the world.

Naturally, the battleships got larger and larger, adding thicker armor, larger cannons, and innovations like revolving turrets—first featured during the Civil War on the *Monitor*. They also became more and more expensive, with massive dreadnoughts costing England and Germany upward of one fourth to one fifth of their entire military budget. The naval arms race—for more powerful guns, thicker armor, and faster engines—pushed naval engineers to the limit of their expertise and governments to the limits of their treasuries. Yet the end result was disappointing. When World War I finally erupted in 1914, European battleships were fairly evenly matched, which meant they were never effective at their primary task of overwhelming their opponent. Indeed, the last battleship action in history—the Battle of Jutland fought in World War I—ended in a tactical draw.

The preeminence of the battleship finally ended in the middle of the 20th century, with the arrival of an elegantly simple application of new technology: the airplane. If that big old battleship can't finish off that other battleship, maybe torpedo and bomber planes—which can take off out of range of the battleship and can't be hit by a battleship's big guns—can slip through and do some damage. That is exactly what happened, at Pearl Harbor, the Battle of the Coral Sea, and Midway. And because the airplane could fulfill essentially the same role as naval guns, but at a much farther range and with greater precision, the great powers searched for a way to augment sea power with air power.

The navies of the world needed long-range eyes to reconnoiter the oceans around them, and the ability to strike at the heart of the enemy with long-range aircraft launched from ships at sea. During World War I, individual seaplanes were used exclusively for reconnaissance and anti-submarine duty. They were launched from warships and landed in the ocean after the mission—to be hauled aboard the mother ship by a crane.

Once again, it was the British who took this concept a step further, and built the first ship dedicated to launching airplanes—the aircraft carrier. Other nations followed suit, and by 1939, several world powers possessed aircraft carriers—most notably Britain, Japan, and the United States. The invention of radar further improved the aircraft carrier's effectiveness.

THE ASCENDANCY OF THE AIRCRAFT CARRIER

The surprise attack on Pearl Harbor demonstrated the *power* of the aircraft carrier when Japanese-carrier-launched airplanes decimated the U.S. fleet, sinking or damaging the entire fleet of battleships—but not the American aircraft carriers, which were fortuitously delivering airplanes to remote bases in the Pacific. The Battle of the Coral Sea and the Battle of Midway proved the ascendancy of the aircraft carrier. Midway was the first carrier battle in history and the first naval engagement in which the opposing fleets were never in range of one another—except through the use of their long-range attack planes.

The effectiveness of the carriers was demonstrated by the fact that, as soon as the Japanese lost four carriers to American torpedo bombers off Midway Island, they retreated. Though the Japanese still possessed the *Yamamoto*, the largest and most powerful battleship ever built, and a vast fleet of warships, without carriers the offensive war was over. The formidable *Yamamoto* could not hope to get near enough to the American carriers without first encountering—and being sunk by—carrier-launched aircraft. After Midway, the battleship was forever delegated to a support role, and the carrier became the center of the naval universe.

As World War II progressed, aircraft carriers got larger and larger, and with the American innovation—the steam-powered catapult for launching airplanes—they got more effective, too. Through the use of a catapult, heavier aircraft with larger payloads could be launched. However, larger carriers required a huge level of support to operate effectively. That support came from other vessels—tenders to deliver supplies, destroyers to hunt down enemy subs that endanger the carrier, cruisers to supplement the carrier's own combat air patrols, battleships to protect against surface vessels, and, eventually, radar and sonar vessels to scan the skies and the ocean depths around the carrier and support ships. The carrier task force was born.

DAZZLE CAMOUFLAGE

The boxy, angular plates that sheath the high superstructure on the most recent U.S. Navy warship designs is a type of low observable technology that partially obscures the signature of the vessel. It is a stop-gap measure, retrofitted to older, less stealthy designs and, as such, is the 21st century's equivalent of the surreal-looking "dazzle" camouflage on World War I warships.

Between 1900 and 1914, artists who called themselves Cubists or Vorticists were moving away from representational art and toward abstraction—stressing forms, shapes, and colors as a means of representing an object on two-dimensional canvas. This artistic philosophy was put to practical use with the "dazzle" effects painted on the sides of warships. A jumbled scheme of surreal, jagged, chopped lines of light and dark primary colors, dazzle camouflage was meant to distort perspective to fool observers—mainly German U-boat captains peering through periscopes. Supposedly, dazzle distorted the shape, size, and direction a warship was moving.

The stealthy, non-radar-reflective superstructure on modern Navy vessels are only marginally more effective than "dazzle," but they help to mask the ship from some radar bands.

After the war, jets replaced piston-engine aircraft, and diesel power was replaced by nuclear energy, but the task and effectiveness of the carrier remained the same. The British more or less abandoned the race to build larger aircraft carriers and relied on less effective, catapult-less carriers to launch vertical take-off and landing (VTOL) aircraft like the Harrier. The Soviets tried innovative designs like the sloped deck and the catapult, but basically failed. As the world's remaining superpower, the United States is also the only nation that successfully deploys carriers worldwide. In the 18th, 19th, and early 20th century, the British Empire kept the sea lanes safe for commerce. Today, the United States has supplanted the British in performing this critical role, and the carrier has supplanted the gunship as the instrument of enforcement.

Now, at the beginning of the 21st century, the carrier task force is the keystone of American defense policy. Yet of all the weapons systems in the world today, the carrier is the one most close to final obsolescence. In fact, the first gong of the carrier's death knell was sounded more than 40 years ago.

During the Cold War, the NATO plan in the event of a conflict against the Soviet Union was to keep the Atlantic sea lanes open for the movement of troops and supplies. This could only be accomplished by keeping Soviet land-based, long-range fighters and bombers, and Russian submarines, away from those convoys. U.S. and British carriers were crucial for this task.

In the scramble for ways to sink NATO ships—yet still remain out of range of the carrier aircraft or air defenses—the Soviets developed the AS-1, its code name, Kennel, in the late 1950s. The AS-1 was essentially a stand-off-and-shoot, airborne-launched MIG-15 fighter that was strapped to the wing of a Backfire bomber. With an explosive warhead replacing the pilot, the AS-1 was aimed by the crew in the bomber that launched it. Essentially, it was a dumb bomb, for the AS-1 lacked a guidance system.

The AS-2 Kipper was a marked improvement. It had better range and speed—up to Mach 1.5—and the pilot who launched it had some limited control over its trajectory even after the cruise missile was launched. The AS-3 didn't survive the research stage, but the AS-4 Kitchen missile was a lethal new breakthrough for the Soviets. It had a terminal guidance system, a range of more than 200 miles, and a rocket engine that boosted it to speeds of Mach 3.5. It could fly high and then dive toward its target, making the Kitchen very difficult to kill. Throughout the 1970s and 1980s, Soviet technology improved, and the AS-6 Kingfish, circa the mid-1980s, was capable of supersonic speed and had passive and active radar that could lock onto a target after the cruise missile had been launched.

The Soviet emphasis on developing heavy, long-range precision-guided cruise missiles was a threat to U.S. naval strategy, which relied on a carrier task force that was now very vulnerable to these new cruise missiles. The U.S. Navy's answer to Soviet cruise missiles was the development of an integrated radar and defensive shield to protect its carrier task forces. The first program to develop such a shield was the U.S. Navy's *Typhoon* system, but it was aborted in 1963 because the technology to make it work had not yet been invented (this refrain sounds hauntingly familiar to proponents of the Star Wars Defense System). But Typhoon was revived in the 1980s and given a new name—*Aegis*, after the shield that Zeus presented to Athena.

The key to Aegis' success is systems integration, which is made possible by the development of high-speed computers that did not exist in the 1960s. So powerful are the radar and sonar waves emitted by the Aegis system that the arrays could not be placed on the carrier it was meant to protect without interfering with normal carrier operations. A whole new platform, dedicated to the deployment of the Aegis system, had to be developed.

The result was the *Ticonderoga*-class cruiser, built exclusively to house the Aegis system. The 27 *Ticonderoga*-class ships are the most complex and capable surface combatants in the U.S. Navy arsenal—in effect, they embody the most recent developments in anti-aircraft and anti-submarine warfare (AAW/ASW). In addition to the Aegis system, they contain 122 Vertical Launch System (VLS) cells that can launch Standard SM-2 MR or ER SAM Tomahawk land-attack cruise missiles, as well as ASROC ASW missiles (see Chapter 3). They can also launch Harpoon anti-ship missiles and torpedoes. Most impressive are the twin Phalanx rapid-fire close-in weapons system (CIWS) designed as the final defense against cruise missiles. The Phalanx is able to fire up to a thousand 20mm rounds of ammunition per second. This sets up a gauntlet of projectiles a cruise missile must negotiate to get through to the carriers.

The Aegis system can be divided into three parts: the sensors, the battle management system, and the weapons. *Ticonderoga*-class vessels carry three different types of surface-search radar, and an AN/SLQ-32 system designed to identify and track airborne targets in an environment jammed with electronic countermeasures. There are three sonar systems for underwater detection, including a hull-mounted and towed-array radar system. All of these systems feed data to the command decision system (CDS) that helps the CINC (Commander-in-Chief) assess threats, assign priorities, and task the weapons, which are in turn controlled by the computerized weapons control system (WCS). The data comes in four modes:

- **Automatic,** in which raw data is fed directly to CDS, then to the WCS, and then to the individual weapon
- **Special Automatic,** in which human controllers can preset priorities for targets
- **Semiautomatic,** in which human controllers interface with the automatic systems
- **Casualty Automatic,** in which systems reroute data and weapons as systems are knocked out of action

Both the *Ticonderoga*-class and the smaller, cheaper *Arleigh Burke*–class destroyers also designed to house the Aegis are phenomenally expensive to build, at more than $1 billion each. They carry a crew of 350 (*Ticonderoga*) or 308 (*Arleigh Burke*). They are effective at what they do, but they have shortcomings. For instance, the Aegis cannot operate close to shore without interference from ground clutter, radio, and civilian communications systems, nor can it work two directions at once. For 360-degree protection, a carrier task force needs two of these vessels. And the Aegis system has a saturation point—which is a closely guarded secret. But at some point in an attack, as many contacts rush toward the task force, the data will overwhelm Aegis and one or two missiles will slip through—and a U.S. carrier will die.

The biggest problem with Aegis is that it is basically a defensive system. It adds tremendous cost to the carrier task force, but adds nothing to its offensive capabilities. The same aircraft launch from the carriers with the same payload and fly the same operational range—nothing has changed except that it costs a whole lot more money just to keep the carriers viable. And with the former Soviets now selling their cruise missile technology to other nations, and with the Chinese entering the race for bigger, faster, larger cruise missiles—it is clear that the aircraft carrier's days as ruler of the seas are numbered. Like the battleship it replaced, the carrier will be rendered obsolete by a simpler, faster, more elegant weapon—the precision-guided cruise missile.

Of course, Navy high command is doing all it can to stave off the carrier's obsolescence and regards the advanced Aegis system as a good first step. There are long-range plans to create new defensive systems—and to build a new type of aircraft carrier.

The apex of carrier construction today is the *Nimitz*-class nuclear-powered carrier, the most recent example of which is the USS *Ronald Reagan* (CVN-76). *Nimitz*-class vessels are based on a 35-year-old Cold War design. The largest warships ever built, with a displacement of 96,700 tons, the USS *Nimitz* (CVN-68) and her sister ships like the *Ronald Reagan* contain 2,000 water-tight compartments, carry nearly 3,000 tons of aviation ordinance, 2.7 million gallons of aviation fuel, and 85 fighter, attack, antisubmarine, and airborne early warning aircraft and helicopters. Since the mid-1990s, some of the *Nimitz*-class vessels shed some of their aircraft to make room for a Marine Corps force of 600 Marines and 10 transport and support helicopters. *Nimitz*-class vessels contain two nuclear fuel cores (early nuclear carriers like the *Enterprise* had eight) and require a crew of 3,200, plus an air-wing crew of 2,800. In one sense, the *Ronald Reagan* will be the last of her kind because another vessel—the not-yet-named CVN-77—will be the last *Nimitz*-class carrier ever built, and a *Nimitz* in name only. Though her internal workings and power plant will be the same, the CVN-77 will have very different external features. For one thing, this ship will be designed to accommodate the new Joint Strike Fighter (JSF) to make its debut—Congress willing—in the next decade (see Chapter 11). But the CVN-77 will be unique in that it will serve as an experimental bridge to the next generation of aircraft carriers, the ultra-secret CVX-78 and beyond.

The CVN-77 will embody "low observable" technology that will significantly diminish the carrier's radar, thermal, and electronic signature. This new carrier will also incorporate some of the stealthier design qualities discovered during the joint U.S. Navy/"Skunk Works" experiments with the *Sea Shadow* in the 1990s. Today, due to signature reduction technology, the *Arleigh Burke*–class Aegis Destroyers are extremely tough to see on radar and infrared sensors. The CVN-77, despite its enormous size, may also have a low signature due to structural changes and the use of new "stealthy" materials in the ship's construction.

Other new materials will also be used in the construction of the CVN-77, including heat-resistant silica tiles to protect deck areas from jet blast; lightweight, blown fiber-optical local area network cables that will increase the speed and capacity of the ship's data network; lightweight composites for topside structures to reduce radar signature; and a new generation of nonskid coating that can be applied to just about any substance.

The CVN-77 is also designed to be highly automated to reduce crew size. The first generation of "smart ship" automated technology has been tried with great success on the USS *Yorktown* (CG-47), reducing crew size by 15 percent. On the CVN-77, the Navy plans to reduce crew size further—up to 30 percent—with automated systems replacing manpower at every level of shipboard duty.

Like a modern office building, the CVN-77 will feature rapid reconfiguration technology, for rapidly shifting mission profiles. Interior spaces will feature revolving or fold-back partitions and walls, stowage for large contingents of Marines, along with birthing areas for their air-support wing. The completely redesigned, easily reconfigured interior of the CVN-77 will make the ship more capable for joint operations, so that units like an Army Airborne battalion and all of its helicopters, or a special operations force, can be accommodated. Hangar bays and elevators will be more flexible to accommodate different types of aircraft, from the V-22 Osprey to the new generation of unmanned reconnaissance aircraft.

THE *SEA SHADOW*

The *Sea Shadow* was designed and built by the U.S. Navy and Lockheed as an experimental sea vessel to test stealth technology on the ocean surface. Displacing 560 tons, the *Sea Shadow* was 160 feet long and 70 feet wide. To preserve its secrecy, the ship was built in the early 1980s in stages—eight sections per shipyard, and then assembled "as a gigantic jigsaw puzzle inside a huge submersible barge located in Redwood City, California."[1]

The *Sea Shadow* had a four-man crew, was made of very strong welded steel, and embodied SWATH technology with twin pontoon-shaped hulls. The first radar tests were performed off the coast of Santa Cruz Island in the dead of night. In one test, Navy submarine-hunter aircraft made 57 passes at the *Sea Shadow,* but detected the ship on radar only twice—both times at under a mile-and-a-half distance.

The biggest problem turned out to be the *Sea Shadow*'s invisibility. The ship appeared like a blank spot in the ocean, which made it a dead giveaway. Though the Navy rejected the program because of its "radical design," much of the knowledge gleaned from testing the *Sea Shadow* was integrated into more traditional Navy warships.

The *Sea Shadow*'s design was indeed radical—it looked nothing like a ship. A black, faceted superstructure with sloping sides resembling a barn roof perched on top of two slender submarine-like hulls. From the front it looked like "Darth Vader's helmet."[2] The *Sea Shadow* was propelled by twin diesel engines buried in the superstructure to prevent noise and thermal exhaust. These soundproof engines drove the fanlike propellers via electric motors. The Navy initially rejected the *Sea Shadow* outright in 1993, and the vessel was declassified and used for further research. But the Navy returned to the design in 1999 as a test platform for the technology to be integrated in the design of the *Zumwalt*-class destroyer, which will be the U.S. Navy's first true stealth ship when it is launched.

The schedule for construction of the CVN-77 is near. Funding for the project is to begin in fiscal year 2008, for a possible launch by 2010. The CVN-77 will replace the USS *Kitty Hawk* (CV-63), which is reaching the end of its service life.

The next generation of carrier design has been designated the CVX-78 (Aircraft Carrier—Experimental)—the first vessel in a new class of carriers. Right now the CVX is just a dream of the Carrier Innovation Center, a Navy think tank that, as its name suggests, studies experimental data and new technologies and applies them to new carrier designs. Due to cost and environmental concerns, a new non-nuclear power plant has been envisioned for this vessel, or a new, ultra-efficient nuclear reactor with a conventional electrical generating back-up system, which might be the new, advanced, integrated full electric propulsion (IFEP) system favored by the British for their next generation of carriers.

Hull designs for the CVX-class ships are still being studied, but because the carrier will displace nearly the same amount of water as the CVN-class—approximately 95,000 tons—there are limits to innovation. However, many new technologies—including SWATH-type hull construction that will be discussed later in this chapter—are being considered. Double-hull construction of a different kind—an outside layer of armor, an area filled with reactive foam, followed by the actual hull—is also being considered, though weight and cost are deterrent factors. If the visionaries at the Carrier Innovation Center have their wish, the CVX-78 will no longer employ the half-century-old steam technology to fire its next generation catapult. The high-pressure steam lines take up a lot of room and tend to rupture under combat conditions. Initial designs for the CVX include electromagnetic rail-gun technology, but internal-combustion engines capable of powerful bursts of power are also being considered. The latter would involve a contained fuel-air detonation in a piston that would launch the aircraft from the ship in a simple and reliable system, which would use available jet fuel (instead of the ship's finite electrical power).

The CVX-78 will include a new generation of anti-cruise missile systems, including the new generation of countermeasures against supersonic anti-ship weapons. Designers dream of a carrier design that will embody effective countermeasures against all threats—in essence, embodying the qualities of an *Arleigh Burke*–class destroyer and a carrier. Perhaps the CVX-78 will be that vessel. In any case, if the CVX-78 project stays on track, the first vessel will be commissioned in 2013 and will launch by 2016.

Because an aircraft carrier exists to bring a squadron of attack planes within striking range of the enemy, a few new concepts are being considered for the 21st century. A joint U.S./Kavaerner (a company that manufactures giant oil rigs) venture proposes to build the first Mobile Offshore Base (MOB). The concept gained favor after the Gulf War as a means to place a fighter, bomber, or attack-plane wing on a fixed platform that will serve the same purpose as a mobile aircraft carrier, but much more cheaply.

MOB structures will allow the use of large-capacity transport aircraft such as the Boeing C-17A *Globemaster*, with complete refueling and support services, for re-supply. During a conflict, the MOB can be towed to the theater of operation to provide air support and docking facilities for ship and troop operations without the need—or the political complications—of using land bases. Such structures would be a prime target, but could be defended by Aegis systems, just like a carrier. MOBs would give the United States flexibility and more reliable deployment in international waters far from politically sensitive areas and waffling allies. The MOB proposed by Kavaerner can accommodate up to 10,000 troops and their equipment, and an entire combat wing, including support aircraft and helicopters.

Since the British pioneered the use of smaller, "mini" aircraft carriers in the 1980s and 1990s, several other world and regional powers have entered the carrier business. Small carriers usually operate VTOL (vertical take-off and landing) and STOVL (short take-off vertical landing) aircraft, and that is the scheme behind the carrier operations of Italy, Thailand, and Spain. India maintains a small fleet of used carriers purchased from other nations. Meanwhile, the former Soviet Union struggles to maintain its 40,000-ton carrier *Admiral Kutznetsov*, with its 50 aircraft and catapult-less, ski-launch system.

Strategists in Great Britain, and a new breed of visionary Royal Navy officers, are about to embark on the design and construction of a new generation of aircraft carriers for the United Kingdom. The British CV(F) (Aircraft Carrier—Future) program is running parallel to, and sharing technology and design innovations with, the CVX program in the United States, though on a smaller scale. Future British carriers will displace about half the water of an American *Nimitz*-class vessel—about 40,000 to 45,000 tons—and will no doubt deploy the next generation of VTOL or STOVL aircraft.

The British are also investigating some rather unconventional designs that, if successful, will undoubtedly shape the future of the aircraft carrier. The Royal Navy commissioned the AVPRO company of the United Kingdom to come up with some innovative small carrier designs, and the results were interesting. Many of them embody a breakthrough concept in naval architecture—SWATH technology.

STABILIZING SHIPS WITH SWATH TECHNOLOGY

SWATH (Small Water Plane Area Twin Hull) refers to a relatively new ship type that can greatly reduce instability and wave motion—even in very rough waters—while increasing crew effectiveness and safety. A SWATH ship rides on the water without the large motions and accelerations of conventional vessels. The primary cause of oceangoing instability is the periodic vertical accelerations of single-hulled ships, caused by wave motion on the surface of the water, or "wave action zone."

A conventional V-shaped hull presents considerably more surface area in the wave action zone than a SWATH. By contrast, a SWATH moves the buoyancy of the vessel to the two submarine-shape pontoons that are submerged beneath the waves. Only narrow vertical struts, which present considerably less surface area, are exposed to wave action. The general design of a SWATH includes the two submarine-shape lower hulls that ride submerged, the vertical struts that project above the surface, and a cross structure high above the water to hold the upper tiers of the vessel.

By canceling out wave forces that cause large motions and accelerations, a SWATH ship can offer a level of stability unattainable on a mono-hull or even on a catamaran hull configuration of similar size. The U.S. Navy has several SWATH ships in service now. All of them are research vessels, and none are even as large as a small cruiser. SWATH technology was a major feature of advanced research vessel *Sea Shadow*.

Among the new AVPRO/British designs is the *Air Support Catamaran* (ASCAT), a small vessel to provide limited air support to low-intensity conflicts. Its shallow draft would make it useful in getting close to shore—but still far enough away to avoid shore defenses. Such a weapons platform would be useful in small-scale regional conflicts where the use of a large carrier would be wasteful. The ASCAT would feature a catamaran design, though a version using SWATH technology for added stability has also been considered. The ASCAT could conceivably launch STVOL aircraft, attack helicopters, or unmanned aerial drones. This system would also be useful to humanitarian-aid operations, especially in regions where it is dangerous to place ground forces on a full-time basis.

The *Stealth Trimaran Aircraft Carrier* (STAC) is part of the British aircraft carrier of the future (CV[F]) program, and the UK Defense Evolution Research Agency (DERA) has commissioned the building of a trimaran (three-hulled) research vessel, the RV (Research Vessel) *Triton*. The STAC will feature a low-radar cross-section, making it difficult to detect. The ship will be "stealthy" in design and materials, with sharply reduced radar and sonar emissions. Exhaust from the STAC's engines will be vented through the sides of the inner and outer hull to reduce thermal signature.

The benefit of the STAC concept is that it is adaptable to VTOL, STOVL, and CTOL (conventional take-off and landing) aircraft. As the British see it, the STAC's catapult will not be operated by steam, electromagnetism, or fuel-air explosives. Instead, it will feature a single-use strop mounted on a cylinder that is fired along a tube running below the flight deck. The strop will be shot into the ocean, but will push the aircraft off the flight deck at a rapid speed. The U.S. Navy eschews such wastefulness and prefers a rail-gun configuration.

Perhaps the most advanced design that has been proposed is the *Wing-Assisted Trimaran* (WAT), first developed by an Italian consortium for commercial use. Its naval applications—if implemented—will mark a true revolution in carrier design and tactical deployment. Designed to reach speeds of 200 knots, the WAT is a cross between a hydrofoil and an airplane—or rather, a wing-in-ground (WIG) effect vehicle, because the WAT never completely leaves the surface of the ocean. Rather, it zooms along at 30 knots, but, when required, it can speed up. The design is a trimeran hull configuration supporting a wing. At low speed, the three hulls provide buoyancy. At speeds greater than 35 knots, the wings lift the WAT off the water. Submerged fins provide maneuverability, the propulsion is provided by water-jet engines, and forward-looking radar and sonar provide the eyes. The carrier WAT would deploy a small air wing of up to three strike aircraft. The high speeds at which the WAT moves means that it could launch and recover fixed-wing aircraft from an elevator located in the center of the hull, without the need for catapult assistance. A trio of WATs could provide air control and support for a wide area of operation. Deploying along a coastal region, a WAT or squadron of WATs could respond quickly to rapid movements on the distant battlefield, so that troops ashore would never lack for air support. A fleet of such vessels would be put to good use by the U.S. Marine Corps.

If the aircraft carrier is an endangered species in the shifting conflicts of the 21st century, the nuclear submarines of the Cold War era will soon be extinct. Built to cruise the oceans of the world carrying nuclear-tipped missiles to be fired at the Soviet Union in the event of war, the fall of the Soviet empire has left the U.S. Navy's submarine fleet without a viable, 21st-century mission.

THE NAVAL COMMUNITY

The U.S. Navy is divided into three communities, each with a distinctive mission profile and set of traditions: the submarine service with its nuclear attack and ballistic missile boats; naval aviation with its squadrons and carrier fleets; and the surface navy to escort the carriers, comprised of frigates, destroyers, cruisers, and supply vessels.

The Navy insists there are still many roles and missions that SSNs (Attack Submarines, Nuclear) can support or prosecute, but these missions will tend to be in support of land battles, or as sentries to keep the rest of the world in check while the United States battles a specific enemy. This is the role the SSNs performed in the Gulf War. Though their services were not required in the Persian Gulf, SSNs patrolled the rest of the world's waterways to ensure that no other belligerent power took advantage of the chaos in Iraq to make a preemptive move against another nation.

The 100-plus SSN-fleet envisioned by Ronald Reagan in the 1980s has, quite rightly, fallen by the wayside. Perhaps there is a need for a force of some SSNs, but not that many. In the 21st century, America's submarine force will be shunted aside, to fulfill the types of support and sentry roles the battleship performed before the whole concept of the gunship was sidelined by history.

CHAPTER 8

ALWAYS FIRST TO FIGHT:
THE U.S. MARINE CORPS

The U.S. Marine Corps is older than the nation it serves. Although it's a branch of the U.S. Navy, a Marine is neither soldier nor sailor—he is both. Marines serve aboard seagoing vessels, and wage war on sea *and* land. Today the Marine Corps continues the mission it defined for itself during World War II—that of an expeditionary force trained in amphibious warfare, a force meant to storm a hostile beach and hold it until follow-up forces arrive—in a few hours, days, weeks, or even months.

Traditionally, Marines bring with them what they need, from supplies to armor to artillery, and even their own close air support. In fact, the coordination between warplanes and ground forces was a tactic pioneered by the Marine Corps. This contribution to the art of war is now practiced by virtually every military in the world.

Today the Marines Corps deploys a fleet of ships, a bevy of landing craft, fixed-wing warplanes, vertical take-off and landing (VTOL) aircraft, and attack helicopters specifically designed to deliver troops and their equipment to the shores of battle and provide amphibious invaders with air support and reliable resupply. A new generation of arms, armor, ships, landing craft, warplanes, and aircraft are in various stages of development, which will keep the Marine Corps of the future supplied with the advanced weapons required to maintain the Marine Corps tradition as the first to fight.

SUPPORT FROM THE AIR

In 1919, the 4th Marine Air Squadron was dispatched to Haiti to battle the *cacos* rebels commanded by Charlemagne Peralte. During this conflict, 13 rickety aircraft—outmoded Jn-4 "Jenny" biplanes—and their Marine Corps aviators would become the first to experiment with low-level dive bombing tactics. This was the beginning of modern air-ground operations, first pioneered by Marine Corps aviators like Lieutenant Lawson H. M. Sanderson. The irony was that while the U.S. Navy ignored Lieutenant Sanderson's experiments, both Adolf Hitler's Luftwaffa and the Imperial Japanese navy would utilize many of Sanderson's tactics during the European blitzkrieg and the attack on Pearl Harbor.

Although today's Marine Corps has many missions, it is most associated with amphibious invasions. Marine Corps functions also involve security—for U.S. embassies and diplomatic personnel overseas, aboard U.S. Navy warships, and at nuclear weapons facilities and storage sites—as well as transportation for the president and senior administration. U.S. Marines are also among the most experienced disaster relief and humanitarian-aid workers in the world, with operations in Somalia, Haiti, Kosovo, and Bosnia under their belts.

Since the end of the Cold War, the Marine Corps and the Navy have redefined their roles in the future of war. In this reassessment, the Marines have come up winners. After the Cold War and the end of hostilities in the Gulf War, the Navy published a revised edition of a study, called *Forward from the Sea*, which outlines the Navy's and Marine Corps's mission parameters for the 21st century. In this study, the Navy all but backed away from its traditional role as guardian of the oceans and tied its future operations to the littoral (coastal) regions of the world. Coastal regions in the Middle East, Malaysia, the Philippines, the Indian Ocean, and Asia all are flash points for potential future conflict. With the Soviet fleet gone, the Navy has decided to concentrate on this new "littoral strategy," which will greatly benefit the Marine Corps. Despite drastic-force structure cuts during the

1990s, the Corps has lost less than 11 percent of its former strength because of the accurately perceived importance of their role in future conflicts. Today's Marines are a rapid-strike, shock force—the first to arrive in the theater of battle to secure the beachhead and hold it for follow-on forces. Because the Marines have been the guys who "kick in the door"—in Korea, Vietnam, the Middle East, the Persian Gulf, Somalia, Haiti, and Kosovo—their place at the forefront of American warfare remains secure.

The Marine Corps has done some restructuring to adapt to the threat of small wars. To increase readiness and shorten preparation time, in the 1990s the Marine Corps began to deploy battalion-size landing forces of between 1,500 and 2,500 men to forward bases close to potential trouble spots all over the globe. These *Amphibious Ready Groups* (ARGs) can be deployed anywhere in the world in a matter of days or even hours. Unlike the Army, Marine units travel light—they are basically infantry formations of riflemen, with some air-support elements, and they aren't saddled with heavy armor or artillery. In the past, the Marines enjoyed artillery support from Navy warships off the coast, but the number of surface warships capable of providing fire support are diminishing, so today's Corps relies more than ever on their air support.

Today, Marine Corps units will arrive off the shore of a potential battleground in large amphibious landing vessels. From the flight decks of these massive ships, Marines will deploy in helicopters, even as their air support—VTOL Harriers or Apache attack helicopters—rise from the forward deck, to head for the beach. From the bellies of these same giant vessels, air-cushioned landing craft will roll into the ocean, to skim its surface as they race to shore at 50 miles per hour. Once the Marines secure the beaches, they will continue their advance on foot or in the protection of light armored vehicles (LAVs) that can negotiate rough ocean tides, rivers, or swamps. Well armed with personal weapons, Marines rely on rapid reinforcement by helicopters and landing craft to deliver men and material to sustain their combat activities ashore. In the future, these basic tactics will not change very much. The Marine Corps will continue to deploy a fleet of amphibious warships, landing craft, and aircraft to provide the muscle to storm, capture, and hold hostile beachheads.

By 2010, the Navy plans to deploy a fleet of 36 vessels that will include the massive Landing Helicopter Dockships (LHDs) and Landing Assault Ships (LHAs) and the smaller Landing Ship, Dock (LSD), and LPD (dock ships configured as amphibious squadron flagships). All these types of warships will soon be organized into a dozen amphibious ready groups—ready for rapid deployment—which will then be dispersed around the world. Each of these ready groups will have the capacity to deliver 12 reinforced Marine Corps battalions, comprised of nearly 2,000 troops each, to any combat zone in the world, to perform any type of mission. The new generation of amphibious ships that will transport these troops is larger and more effective than their predecessors. A single LPD helicopter deployment ship will replace four older vessels and perform nearly all of the same tasks.

The first large-scale specialized amphibious vessels built by the Marine Corps were the *Iwo Jima*–class assault carriers. Designed around the hull and engineering plant of the World War II escort carrier, these ships were designed for maximum storage of aircraft, equipment, and troops. For the last 35 years the *Iwo Jima*–class vessels have been in front-line service. Besides serving in several military actions, *Iwo Jima*–class vessels were used to retrieve the *Apollo* space capsules after their moon missions. They also deployed Harrier VTOL aircraft, and performed minesweeping duties in the Persian Gulf War. Now these *Iwo Jima*–class vessels are headed for retirement.

In the late 1970s, as a smaller, cost-effective alternative to the *Iwo Jima*–class vessels, the concept of the Landing Assault Ship (LHA) was born. Dubbed *Tarawa*-class, these warships included a flight deck and a hangar deck for helicopter and VTOL aircraft like the Harrier. With a crew of more than 1,000 sailors and Marines, and troop accommodation spaces for nearly 2,000 troops, the LHAs are—unlike our *Nimitz*-class aircraft carriers (see Chapter 9)—capable of navigating through the locks of the Panama Canal. LHAs are quick to deploy and are adaptable to many theaters, and so can be moved to a mission relatively quickly. They are also heavily armed because they get fairly close to—but still over the horizon (OTH) of—enemy shore defenses. LHAs include the latest available class of surface-to-air missiles, guns, anti-missile and anti-aircraft defensive systems. Five of these *Tarawa*-class vessels were constructed at a cost of more than $1.6 billion each, and served effectively in many combat and humanitarian aid operations. Yet there were not enough LHAs to meet all the needs of the Marine Corps.

It was during the Reagan administration that the keel was laid on the first of what would become the largest and most advanced amphibious vessels ever built, the *Wasp*-class Landing Helicopter Dockship (LHD). Combining features of a traditional landing ship and a close support carrier, the *Wasp*-class amphibious vessels can sustain and support amphibious operations in any type of climate in all littoral regions. They have formidable OTH attack capabilities. This long reach is achieved through the utilization of the Landing Craft, Air-Cushioned (LCAC) (discussed later), CH-53E Super Stallion helicopters, 20 AV-8B Harrier II Vertical/Short Take-Off and Landing-Capable (VSTOL) fighter bombers, and the new MV-22B Osprey—all packed into the mammoth hull of the *Wasp*-class Landing Helicopter Dockship.

The LHD design integrates the cutting-edge 1990s technology in its architecture, materials, and systems. The vessel is compartmentalized and eminently survivable, with defenses against patrol boats and small attack ships—even suicidal bombers—as well as the ability to avoid, withstand, or survive a hit from a mine, a bomb, or even a cruise missile. LHDs can be sealed, and crews are trained to operate in environments contaminated by nuclear fallout, biological weapons, or chemical agents. The flight deck has take-off and landing berths for nine troop-carrying and attack helicopters and 40 Harriers. There is even a portable ski-lift take-off system to be assembled on deck and used in case things get too crowded to launch Harriers from the central deck.

LHDs are armed with two eight-cell Sea Sparrow surface-to-air missiles launchers, and three Mk16 Phalanx close-in weapons system (CIWS) built around the 20mm General Electric Gatling gun, to deal with smaller cruise missiles like the Exocet or Harpoon. The LHAs radars are the SPS-48E three-dimensional search radar; the SPS-49 (V)5 two-dimensional air-search radar; the SPS-64(V)9 navigational radar; a Mk23 Target Acquisition System to detect, locate, and shoot down attack missiles; and an SLQ-32(V)3 electronic warfare system, which includes a bank of four Mk 137 Super Rapid Blooming Chaff launchers that throw up a wall of metal-coated Mylar strips and infrared decoys to blind or confuse incoming missiles. This last flimsy curtain is the final defensive barrier.

LHDs include four operating rooms and a 600-bed hospital. After the construction of the first LHD, the USS *Wasp* (LHD-1), steel construction on the deckhouse was replaced with composite armor to protect the crew from small-caliber bullets and fire damage. Since then, the entire fleet of LHDs has undergone extensive improvements in design and retrofitting of new technology as it is developed.

Below the flight deck and the fixed-wing and rotary aircraft, below the layers of anti-missile and anti-aircraft defenses, deep in the belly of the LHD, are the landing craft that will carry the artillery, armor, ammunition, supplies, and combat vehicles to the beach even as the helicopters (or the MV-22 Ospreys) deposit troops onto the hostile shore. Three landing craft fit into the bow of an LHD, which can open to deploy them right into the ocean. Today, the premier landing craft used by the Marine Corps is the Landing Craft, Air Cushioned, or LCAC.

Today the Navy deploys three classes of landing craft: the low-tech Landing Craft, Utility (LCU); the Landing Craft, Medium (LCM), and the much advanced Landing Craft, Air Cushioned (LCAC), which is quickly edging the other craft out of front-line service. The LCAC has been described as "a pile of Leggo blocks on a flattened inner tube"[1] and this is an apt description of its appearance. But in the military, as elsewhere, appearances can be deceiving, and this ungainly vessel is the fastest, most powerful, most efficient amphibious landing craft ever constructed. The LCAC has brought about the most revolutionary change in amphibious warfare doctrine since the arrival of the helicopter.

Conventional landing craft need optimal tide and surf conditions to make their landings. It is estimated that less than 20 percent of the world's coastlines could be navigated by flat-bottomed, front-ramp landing vessels used in World War II, Korea, and Vietnam. The Marines yearned for a more adaptable and capable vessel, and looked to alternative technologies for a solution. It came in the form of a surface-effect vehicle—a hovercraft.

MARINE CORPS LANDING CRAFT: A SHORT HISTORY

When Marines hit the beaches in World War II, they did it in boats designed for amphibious operations. The first was the Eureka, built by Andrew Higgins in New Orleans. The Higgins boat had a flat bottom and a propeller protected by a tunnel, so it could run onto beaches, reverse torque, and back away again. In 1941, the Marines showed Higgins a photograph of a boat developed by the Japanese. It looked like the Higgins, but had a forward ramp that dropped to unload. Higgins added a ramp to his design and the LCPR—Landing Craft, Personnel (Ramp)—was born.

Next came the LCM—Landing Craft, Mechanized. The predecessor of the LCM was developed as a rescue vehicle by Donald Roebling, grandson of the man who built the Brooklyn Bridge. The Marines turned this vehicle into the "Alligator," the first LVT—Landing Vehicle, Tracked. Add a gun turret and armor and the result was the "Buffalo." Another Marine Corps vehicle came courtesy of the United States Army. The DUKW—pronounced *duck*—was basically a truck that could swim. The most popular was a 2.5-ton model that was used from 1944 through the end of the war.

Hovercrafts float on a cushion of air contained by a flexible "skirt," and can move over almost any type of surface because it barely touches it, but rides above the surface on a pillow of trapped air. Because the air cushion is virtually frictionless, very little forward thrust is needed to propel a hovercraft—even at high speeds. Hovercrafts are immune to rough weather and rough seas, and can easily transition from sea to land—moving across the ocean, up the beach, along the sand, and onto land.

The Soviet Union worked to develop economical and efficient surface-effects military vehicles in the Caspian Sea and along its marshy shores during the Cold War (more on this later), but in the West it was the British who adapted this technology for military use. British-designed, American-built (by Bell) SR.N5s saw service in Malaysia and Vietnam.

In the late 1970s, the Navy first tested this technology, and it found general acceptance in the Marine Corps. By the late 1980s, the first LCACs saw service. Seventeen LCACs deployed from six LSDs saw service in the first Gulf War, where the vehicle achieved a near 100 percent reliability rate. LCACs have also been used in peacekeeping duties in Somalia and Haiti and humanitarian aid missions after the devastating 1991 cyclone in Bangladesh.

The LCAC serves as the Marine Corps's premier landing craft, with 90 of them currently in service. The Japanese military is also using its own specialized versions of the American LCACs.

Depending on the class and configuration of the "mother ship," from one to four LCACs (usually three) can be stored in the bows of an LHD, an LHA, or an LSD, which has an internal docking facility for mooring landing craft inside its massive hull. All these warships flood a portion of their bows and open a ramp to release the LCACs. It seems incredible that up to three LCACs can fit into the hull of an LHD because the LCAC itself is so large—many times larger than the landing craft used to hit the beaches of Iwo Jima, Okinawa, or Normandy in World War II. Today's LCAC has a 27-foot-wide cargo deck with a length of 88 feet and interior crew and troop quarters forward (in front of its tandem engines). The LCAC has the capacity to carry the same basic payload of an older type LCM (Landing craft, medium)—but at more than four times the speed, over rougher water, over obstacles up to four feet high, and across the beach. Even on land, the LCAC can clear obstacles up to four feet high, and negotiate trenches, anti-tank, and barbed wire barriers.

The cockpit of an LCAC is located on the deck, in a pilothouse on the starboard side. The pilot's and navigator's control station more resembles an airplane flight deck than the bridge of a seagoing vessel, as befits a vehicle that is more helicopter than boat. The LCAC has a crew of five, the most senior of which is the craftmaster. LCAC payloads weigh in at around 75 tons, which could include a mix of personnel and/or vehicles. An LCAC can transport a single M1A1 Abrams Main Battle Tank, or five light armored vehicles, or two M198 towed howitzers and their movers plus two dozen combat-ready troops. During the Gulf War, as many as 41 Marines were deployed from a single LCACs.

LCACs are girded with a drooping black rubber skirt that is inflated by the forward engine unit on each side. Textron Lycoming Gas Turbines drive four lift-fans aimed at the surface water or ground, while the twin after-engines propel the LCAC forward at speeds of up to 50 knots, using 12-foot, 4-blade, reversible propellers housed in an inflatable, collapsible rubber shroud. When the skirt is deflated, the LCAC sinks to the ground so the forward ramp can be deployed for boarding or unloading of troops and vehicles.

During the Cold War, the former Soviet Union tried to develop hovercraft far larger and more advanced than the Marine Corps LCAC. The Soviet *ekranoplan* was designed to carry hundreds of troops into battle over long distances and at very high speeds. The KM8 *ekranoplan*, dubbed the "Caspian Sea Monster," was literally a boat with two stubby wings to lift it up to 50 feet above the surface of the waves. Catering to the Russian propensity for all things gigantic, the KM8—at 458 feet long and 540 tons—was larger than a B-52 and heavier than a 747 jumbo jet. The Soviets built three of these formidable behemoths, two of which crashed in bad weather. Despite the loss, the Russians made plans to build many more of these *ekranoplan* surface-effect vessels, until the Soviet Union crumbled.

Today, both British designers at Avpro, United Kingdom, and American researchers at Lockheed-Martin are working to build a new generation of smaller, faster, more efficient *ekranoplan*-type surface-effect, or ground-effect vehicles (as the British call them). The British beat the Americans into the design phase with two new projects, the Marauder OGE and the Manta IGE.

The Marauder OGE, or "Out of Ground Effect" vehicle, is the smaller of the two proposed designs. The OGE allows the Marauder to actually fly like a conventional aircraft as well as skim the surface of a body of water. The twin-turbofan-powered airframe will be built of composite materials for stealthy characteristics, and will be modular in design so that the basic airframe can be adapted to many oceangoing roles, from aircraft early warning to anti-submarine warfare, ground attack and close-air support, search and rescue, as a landing craft for special operations forces (Navy SEALs or Special Air Service), or for infiltration by a small exploratory force of a dozen Marines (Royal or U.S.).

The Manta is a far more ambitious project, and although it is an IGE (In Ground Effect) vehicle—which means it cannot leave the surface of the water and fly like the Marauder—the Manta is far larger, and is more adaptable for future combat, too. The Manta will be capable of carrying a payload nearly the size of a U.S. Air Force C-174 Globemaster III transport aircraft or an LCAC—about 75 to 77 tons—at speeds of up to 350 miles an hour. Its cargo bay will be 50 feet wide, 46 feet high, and 150 feet deep.

Using a skirt to contain the air, the Manta will hover about 15 feet above the surface, and will be capable of running onto land, though its primary mission will be to cross water, ice flows, marshes, and swamps, even in the roughest weather. The Manta can be ship-based like a helicopter or an LCAC, and one design-configuration of the Manta will feature a helicopter launching platform on its hull. This high-speed vessel would serve excellently as a landing craft or maritime transport, or for search-and-rescue missions. And—like the Marauder—civilian uses like oil-spill cleanup and disaster relief are also being proposed for this unique new class of transports.

Though the golden age of the flying boat supposedly ended in World War II, a new generation of stealthy, high-tech flying boats and ground-effect craft—all based on the Soviet *ekranoplan* design—will make their debut in the coastal regions of the 21st century.

Since the Vietnam War, when the helicopter was introduced to combat, it has become the primary means of transporting troops onto the battlefield. The Marine Corps has used helicopters like the Sea Stallion and Sea Knight, but have generally preferred water-borne landing craft because they can haul a larger payload and are far less vulnerable than choppers are to enemy fire as they approach a hostile shore.

But the next generation of Marines will probably ride into battle in a unique new aircraft—a combination rotary and fixed-wing aircraft called the Osprey. The program to develop the Osprey has been troubled—multiple crashes, not related to the aircraft's designs, have delayed the program, and it was actually cancelled during the 1980s. But that left the Marine Corps without a helicopter to replace the 40-year-old Sea Knight, so the Osprey program was revived in the 1990s.

At $32 million a copy, the Osprey is phenomenally expensive, but not in relation to the tasks it can accomplish. The Marine Corps variant, the MV-22, is just reaching active service now, with its first squadron deployed. The Osprey is about 57 feet long, with a wingspan of 50 feet. It is about 23 feet high, and weighs in at 60,500 pounds when fueled. Performance will include a level flight capability of 315 miles per hour, a "dash speed" of 370 miles per hour, and a ceiling of 30,000 feet. The MV-22 Marine Corps variant of the Osprey can ferry 24 fully loaded troops or 20,000 pounds of equipment. The two-man cockpit of the Osprey is based on the design of the MH-53J Pave Low III SPECOPS helicopter and the MC-130H Combat Talon II aircraft, with a control stick, left-side thrust-control lever and flat, multi-function display panels.

What is unique about the Osprey is the way it takes off and flies. It's a Vertical/Short Take-off and Landing (VSTOL) aircraft with a tilt-rotor. The engines at the end of each wing can elevate 90 degrees for vertical take-off, and then tilt back to full vertical for forward flight. Intermediate angles may be selected. The pivoting Allison T406-AD-400 turboprops drive large, three-bladed graphite-epoxy rotors in opposite rotation to provide both life and thrust. The forward-swept, composite-material wing is carried on a titanium ring to allow it to transverse 90 degrees to align with the fuselage for storage on aircraft carriers or amphibious vessels like the LHD.

Several stealthy characteristics are built into the Osprey, including infrared suppressors on the engines. The flight deck is maintained at slight overpressure, to protect the crew from nuclear, biological, or chemical (NBC) attack. The flight deck features side-by-side seating, and a rear ramp for rapid deployment. Cargo can also be slung from twin external hooks, and the Osprey sports retractable landing gear. As a safety feature, the MV-22 has a built-in Vibration, Structural Life, and Engine Diagnostic (VSLED) system. Pilots can use the Osprey in conjunction with advanced display systems that include night-vision goggles (NVG), heads-up display (HUD), and Forward-Looking Infrared Radar (FLIR) output. Proposed systems on later models will include a gun turret slaved to the pilot's helmet and display. The Osprey also features electronic and anti-missile countermeasures, the details of which remain classified.

Once the MV-22 Osprey enters full service by 2005, it will replace the outmoded Sea Knight and Sea King helicopter still in Marine Corps service. The Osprey is a complex, highly advanced solution to the problem of troop transport.

Marines also reach the beach by using light armored vehicles (LAVs). Descendants of the primitive, reef-climbing Water Buffalo and the other "Amtracks" (Amphibious Tractors) used in the Pacific Theater of World War II, today's light armored vehicles came into Marine Corps service in the 1980s to fulfill the role of a general-purpose armored vehicle for mobile warfare. Unlike the Army's M2 and M3 Bradley infantry fighting vehicles, the Marine Corps armored vehicle had to be amphibious, lighter than the Bradley, and narrower to fit into confined spaces like ships and on LCACs. Although it is neither as capable nor as sophisticated as the Bradley, the basic LAV is far cheaper, and more adaptable for the Marine Corps specialized usage.

One way the LAV was lightened was by replacing the Bradley's tracks with oversized wheels—eight of them. The LAV can negotiate many types of terrain with these wheels, and can be converted to amphibious use in three minutes. All LAVs are armed with a 7.62mm machine gun and eight smoke grenade launchers. Fighting versions like the LAV-25 have a fully revolving turret sporting a M242, 25mm Bushmaster Cannon. The LAV-AT (Light Armored Vehicle/Anti-Tank) sports a TOW (Tube-Launched, Optically-Tracked, wire-guided) precision-guided missile launcher in place of the Bushmaster. There are also supply versions, mortar variants, recovery vehicles, and command versions of the LAV.

A larger, heavier, tracked version of the amphibious armored personnel carrier has also been developed for the Marine Corps. The United Defense AAV7A1 (Landing Vehicle, Tracked, Personnel) is a full-tracked, heavily armored amphibious vehicle capable of over-the-beach landing of up to 25 Marines. Propulsion on the water is through twin water jets. Steering on water is accomplished via a deflector stream at the rear of each water jet.

The AAV7A1 has many drawbacks. Its high silhouette makes it an easy target. It is slow and somewhat cumbersome on water. And it is a little too large for the Marine Corps's liking. In any case, plans are underway to replace it with a more advanced, highly specialized new design for the 21st century.

The Advanced Amphibious Assault Vehicle (AAAV) will be built by General Dynamics, Land Systems, and will embody the cutting-edge of amphibious warfare technology. This AAAV is designed to move at more than 25 knots on water, with sufficient range so that the ship that launches it can remain over the horizon. Its hull is based on "skimming bricks" technology using the basic bow shape of a surfboard to achieve relatively high speeds. To help improve speed in the water, the tracks and wheels actually retract into the hull to reduce drag.

The dual-mode propulsion system will be powered by a 2,600 horse-power MTU/Detroit Diesel turbo-charged diesel engine that is sealed as a self-contained unit. The engine is built to be virtually maintenance free, lasting up to nine years and requiring an oil change every two years. On the water, the AAAV will move at 45 miles per hour, and once it hits the beach it will move faster than an M1A1 Abrams tank—up to 60 miles per hour. New advanced composites created by United Defense will keep the vehicle light, at around 36 tons.

The AAAV will have an electronics package ideal for the digital battle-field, with the same basic electronics package on the advanced Abrams tank, including FLIR viewers for the gunner and driver, GPS slaved to a map screen, onboard diagnostic systems, and SINCGARS jam-resistant radios. A fail-safe system will right the AAAV if it is capsized by enemy action or rough seas. It will be capable of transporting a 13-man rifle squad, their supplies and equipment, with a range up to 300 miles.

Currently, AAAV is designed with a Bushmaster cannon, but that may be replaced down the road by a precision-guided munitions (PGMs) like the third-generation TOWs now in the design stage. The Marine Corps plans to field more than a thousand AAAVs when they all reach service by 2012. The Advanced Amphibious Assault Vehicle will likely be the last armored vehicle produced by the Corps for the next 30 years, and so this weapons platform should have a long and successful life.

Because of the retirement of many of the U.S. Navy's older cruisers, off-shore fire support for future Marine Corps amphibious landings on hostile beaches has been severely curtailed. The Navy has reduced its gunship fleet considerably due to their nearly complete obsolescence. But one thing those big guns were good for was throwing huge shells 20 miles to hit the enemy emplacements on the coastline—true OTH combat capacity. While attack helicopters and Harriers can take up some of the slack, there is a need for off-shore fire support in the future of amphibious warfare.

The solution is the Arsenal Ship, a concept not seen in the U.S. Navy since the riverine battles of the American Civil War. Like the gunboats of old, these arsenal ships will be simple and relatively inexpensive vessels with missile launchers packed with up to 700 tactical missiles, including Tomahawks or perhaps the next-generation of supersonic, precision-guided cruise missiles. Future arsenal ships are intended to be one-shot wonders—they will fire their payload in a single salvo, and then return to harbor to reload. The warship would rely on off-board targeting—by drones, manned aircraft, GPS-transmitted data, special operations forces painting the target from the ground, or simply advanced intelligence and pre-programmed navigational packages installed especially for that particular conflict.

Constructed of composite material and covered with radar-absorbing, nonreflecting coating, the Arsenal Ship would have virtually no super-structure, which would make it quite stealthy. There are even competing Arsenal Ship designs that include partial submersion capabilities to further elude enemy detection.

All this planning, all this innovation, proves one thing—there is a continuing role for the Marine Corps in the future of American warfare. It leads one to imagine that on some very distant planet, heavily armed and armored troops in environmental gear will drop out of high orbit in reentry ships, to seize a piece of ground from the enemy and hold it until reinforcements arrive. Whoever these men are, they will be part of a long and honorable tradition, for they, too, will be Marines.

PART 4

THE HIGH FRONTIER: AIR WAR IN THE 21ST CENTURY

"To conquer the command of the air means victory; to be beaten in the air means defeat."
—Giulio Douhet, *The Command of the Air*, 1921

"Modern air power has made the battlefield irrelevant."
—Sir John Slessor, *Strategy for the West*, 1954

CHAPTER 9

THE FUTURE OF AIR COMBAT

When the first primitive aircraft appeared in the skies over the Western Front, these ungainly wooden, wire, and canvas contraptions represented the cutting-edge of military technology. Though the first airplanes had little operational impact on the outcome of World War I, visionary thinkers recognized the potential of the warplane in the future of war. As early as the American Civil War, and later in the Franco-Prussian War of 1871, airborne reconnaissance from hot-air balloons was tested in a limited fashion. But the first true demonstration of air power's *lethal* potential came in 1911, when Italian army lieutenant Giullio Gavotti made the first bombing run from a heavier-than-air platform, using his early model biplane to launch a grenade attack against Libyan troops camping in an oasis. Though the impact on the Libyans was marginal, Gavotti's attack demonstrated the potential of the bomber—an airborne weapons platform that could rain destruction upon the enemy.

During World War I, the bomber was introduced with limited success because technology was limited. Both heavier-than-air warplanes (such as the German Gotha and Zeppelin-Staaken bombers) and lighter-than-air craft (German zeppelins) attacked London and Paris with aerial bombs, inflicting little damage. The militaries of the world longed for an improved weapons platform, a bomber that could fly without restriction over rough terrain, through anti-aircraft defenses, and over concentrations of enemy forces to strike at strategic targets far from the front lines.

Between the two world wars, theoreticians like the American Billy Mitchell and Italian General Giulio Douhet instantly grasped the vast potential of air power. It was the Italian general who first codified strategies and tactics to successfully wield air power. Douhet deduced that because modern warfare relied on the industrial base of an entire nation to support it, by destroying that infrastructure you destroy the ability of a hostile power to wage war. In Douhet's view, the factory, the civilian factory workers, and the civilian social systems to support those workers were legitimate military targets because they were essential to the production of war materials. Without rifles, bullets, tanks, or airplanes, the enemy could not sustain hostilities and must surrender. Douhet believed that aerial bombardment would also be a potent psychological weapon, one that would demoralize the civilian population and lead to social chaos.

Back in 1921, many questioned how such destruction was to be accomplished given the low level of aviation technology, and the general didn't help his case by speaking in absolutes. In *The Command of the Air*, General Douhet insisted that the guiding principle behind all aerial bombardment should be the goal of destroying the target completely in a single attack, making further attacks on the same target unnecessary. Yet even at the outbreak of World War II, two decades after the publication of Douhet's groundbreaking work, this was an impossible goal. Airplanes couldn't fly far enough, or carry enough bombs, or aim them accurately enough to destroy a target in a series of raids, let alone in a single strike.

THE FAILURE OF PEARL HARBOR?

According to Douhet's principles, even the Japanese attack on Pearl Harbor, one of the most successful single air attacks in history, was a failure because the harbor facilities remained intact and continued to service U.S. Navy warships throughout the war. Thus, the Japanese failed to destroy the target in a single decisive strike!

Though the airplane had evolved into an effective and successful weapon by the outbreak of World War II, the search for the perfect bomber continued unabated throughout the war and beyond. Four critical criteria must be considered in the design of an effective bomber aircraft: range, payload, survivability, and accuracy. The farther a bomber can fly, the more successful it will be, because aircraft that can fly an adequate *range* will be capable of fulfilling more combat roles. Long-range bombers can be used to strike at the enemy's infrastructure, enemy supply lines, and concentrations; or they can be used to patrol coastal areas and the vast reaches of the ocean to protect friendly convoys while at the same time hunting for enemy submarines and surface vessels.

With good range, a bomber must also possess an adequate *payload* capacity—it must be able to carry enough bombs to inflict damage on a target. In designing a conventional bomber, the struggle between range and payload is constant—an increase in range usually means a diminishment of payload. Today's high-speed jets have solved many of the payload problems that plagued propeller-driven fixed-wing aircraft, because an aircraft that can travel at very high speeds can carry more stuff. But during World War II, an increase in payload capacity automatically meant an increase in engine power to lift it, but more engines meant more weight, which further diminished the payload. The right combination of power and payload was a balancing act that aircraft designers seldom achieved to perfection.

Yet more than range and payload are required of a good bomber. It must also be capable of *surviving* the incredibly hostile skies over enemy territory. Radar, anti-aircraft guns, and fighter-interceptors all conspire to stop a bomber from reaching its intended target or escaping the area after its bombing run. Speed, maneuverability, and external defenses all increase survivability—and all present problems. Bombers are not generally known for their speed—the supersonic bomber is a relatively recent invention—nor are they famed for their maneuverability. Bombers in World War II tended to rely on external defenses. The problem is that machine guns and gunners, turrets bristling with weapons, and anti-radar chaff dispensers (invented in World War II) tend to add weight to the aircraft, which means diminished payload, more engines, and further complications to the endless cycle of range vs. payload.

The final, and perhaps most important, requirement for an effective bomber is *accuracy*. A bomber must be able to locate and hit its target to accomplish its mission. Range and payload and survivability are nothing without accuracy, and building an accurate bomber became the most difficult problem of them all to solve. Though every bomber built during World War II possessed two or three of these characteristics, no bomber built by any combatant in this war possessed them *all*.

Because the militaries of the world lacked a bomber that combined long range, large payload, survivability, and accuracy, the basic promise of air power to be a decisive factor in warfare was never truly realized in the 20th century. General Douhet would have to wait until technology caught up to his pure vision of dominance through air power. It would be a long wait.

During World War II, General Douhet's philosophy was put into practice, pushed by forward-thinking tacticians like "Bomber" Harris and Curtis LeMay. But from the start, conventional high-altitude bombing never lived up to its promise. During the blitz, London was bombed nightly, and then hit with V-1 and V-2 terror weapons, yet the basic infrastructure continued to function, and the social order never collapsed. In the latter part of the war, both Berlin and Tokyo were bombed, yet they also continued to function without the predicted collapse of their respective societies or an appreciable diminishment of wartime production.

As enemy defenses improved, British bombers like the Lancaster suffered heavy losses and were deemed too vulnerable for daylight raids. To protect their bombers and crews, the British began a campaign of high-altitude nighttime bombing, despite the fact that high-altitude *daylight* bombing—where the target was usually clearly visible to the naked eye—had proved to be wildly inaccurate. Poor navigation, inefficient bombsights, wobbly bombs, wind drift, fighter attacks, and antiaircraft fire combined to render accurate bombing next to impossible. After the war, it was discovered that on nighttime missions, most British bombers were lucky if they dropped their bombs within *three miles* of the target.

The U.S. Army Air Corps firmly believed in daylight bombing and stuck to it, even as it sought to improve the survivability and accuracy of its bombers in a noble attempt to limit what we now call "collateral damage." The quest for survivability led to the ascendance of the rugged Boeing B-17 "Flying Fortress." As its name implies, the B-17 was a four-engine, heavily armed long-range bomber packed with defensive machine guns in its waist, nose, belly, and tail—literally a fort with wings. A boxed formation—what the Luftwaffa pilots called a "Pulk"—was developed by the Eighth Air Force to prevent enemy fighters from penetrating the B-17 formations without first running through a gauntlet of up to 40 or 50 machine guns. Despite these factors, the daylight bombing runs over Germany were terribly costly in both men and material and did not have a decisive effect on the production of war materials.

The quest to improve bombing accuracy led to the development of the Norden bombsight. Essentially, an automatic pilot that took control of the aircraft, stabilized it through a series of gyroscopes, and adjusted for trim, altitude, and wind velocities. The Norden bombsight was the best aiming device of its kind in World War II. It also didn't work very well. The Norden could not pierce cloud cover, smoke, or ground fog, and was not capable of directing the bombs—only the bomber.

Despite widespread use of the Norden system, accuracy was hardly improved after many months of bombing, and Allied losses of bombers and crew remained very high. In desperation, the Army Air Corps resorted to saturation bombing, which made a certain amount of tactical sense. If the bombers couldn't hit the target with precision bombing, perhaps saturation of entire areas with hundreds of bombs would do the trick. It wasn't long before saturation bombing with conventional high-explosives led to saturation bombing using incendiary devices such as napalm. Incendiary bombs were used with horrific consequences on Dresden, and later in Tokyo. The perceived effectiveness of saturation bombing ultimately led to the use of the first atomic weapons on Hiroshima and Nagasaki—saturation bombing of another kind.

After the two atomic attacks, the Japanese promptly surrendered. It seemed as if air power finally worked as Douhet envisioned it would; yet appearances were deceiving. In the Pacific, the vast distances between the Allied and Japanese bases made strategic bombing impossible for both sides. Early on, the air war against the Japanese was conducted almost exclusively by carrier-borne aircraft. After the Marine Corps spent two years capturing island after island in bloody amphibious invasions, the Air Corps finally had land bases close enough to the Japanese mainland to launch bombers to strike at cities and factories. Yet despite the grave sacrifices made to secure these bases, conventional bombing and saturation bombing failed to break the will of the Japanese people or to curtail their wartime production.

Ultimately, the United States used air-dropped nuclear bombs, and the terrifying effect of these weapons compelled Japan to surrender. But these weapons were carried in two B-29s that took off from a forward base on Saipan, a base taken from the Japanese by the Marine Corps. Without that forward base and others like it, long-range bombing of Japan would have been impossible. And it should be remembered that there was no way that air power alone could have captured those islands. It took Marines, landing craft, artillery, tanks, and flame throwers—not precision bombing—to hop from island to island, moving ever closer to Japan.

This is not to say that air power was utterly impotent. The bombing campaigns in Europe and Japan had an impact on the outcome of the war. And the virtues of close air support, in the form of dive bombers, torpedo planes, and attack fighters, was proven beyond a shadow of a doubt. But the overwhelming superiority of aerial bombardment over all other means of waging war—the core of General Douhet's philosophy—had not been proven by the events of World War II.

Nowhere was the *apparent* failure of air power more evident than in the Vietnam War; but, again, remember that looks can be deceiving. In Vietnam, the United States fought a purely defensive war, which is almost always a losing proposition. More disastrous than that, U.S. policy in Southeast Asia was illogical: America was dedicated to halting the North Vietnamese from invading the South *without decisively defeating the North*. This was because the memories of the Korean War were still fresh, and the administrations of Kennedy, Johnson, and Nixon all feared that if North Vietnam was threatened with annihilation, then the Chinese would enter the conflict as they had in Korea.

Unfortunately, the U.S. government also chose to fight the Vietnam War on the cheap, at least initially. America was slow to move large amounts of troops to the region or call up reserves, and instead relied on South Vietnam to mount its own defense against the North and to do much of the fighting. Even worse, the Kennedy and Johnson administrations—guided by Secretary of Defense Robert McNamara— also sought to use strategic bombing as a wedge to force the North to capitulate. Their rationale came right out of Douhet's *The Command of the Air:* Destroy North Vietnam's infrastructure, sow social chaos, and the Communists would have no choice but to give up the struggle to unify Vietnam and retreat to the North. But this logic was inherently flawed, as the North Vietnamese *had* no infrastructure. They were a primitive agrarian culture, their dominant product rice. It was impossible to bomb the North Vietnamese back to the Stone Age—they were already there. The factories that built the weapons that supplied the North were located outside Moscow, Beijing, Prague, and Pyongyang, not Hanoi. The United States was not willing to bomb these cities and risk a world war, nor were they willing to interdict the ships that brought these weapons to the North Vietnamese through Haiphong Harbor.

Taking these factors into consideration, it is difficult to see how unrestricted bombing of North Vietnam *could* have defeated the Communists—and the bombing campaign in Vietnam was anything but unrestricted. International political pressure kept the U.S. Air Force from hammering vital strategic targets. Political considerations had precedence over military considerations, and the North Vietnamese soon learned which targets would be hit and which would be spared. For instance, virtually all of the war supplies entered North Vietnam through Haiphong Harbor, or on two rail lines emanating in The People's Republic of China. Yet all three of these targets were off-limits to U.S. bombers. The U.S. government feared that bombing the harbor would have international political ramifications, and hitting the rail lines might cause the Chinese to react militarily.

So the Air Force was compelled by circumstance to focus its attentions elsewhere. An effective and deadly air and ground campaign to interdict supplies coming south along the Ho Chi Minh Trail or through

Laos was launched. Though the trail had no choke points and shifted constantly in response to American military actions, thousands of tons of material and thousands more North Vietnamese regulars were destroyed by U.S. ground troops, low-flying attack planes, and unique developments like the C-130 gunship, designed specifically for interdiction duty. Of course, the Americans could not hope to interdict all of the supplies coming from the north. There were too many trails, too many ways to move supplies, and, once again, political considerations restricted all-out action.

U.S. Navy carrier-based aircraft and the Air Force also concentrated their attention on the nexus of the Chinese rail lines—the Paul Doumer Bridge, built by the French to span the Red River near Hanoi, and the Thanh Hoa Bridge, which crossed the Chu River. War materials crossed these two railroad bridges, to arrive in Vinh, where they were off-loaded and brought into the south or through Laos along the Ho Chi Minh Trail. These bridges were located in North Vietnam and were deemed legitimate targets. They were heavily defended by anti-aircraft fire, and, as the Americans would soon learn, by Russian MiGs flown by Soviet "advisors."

Beginning in April 1965, the Air Force and Navy launched repeated air strikes in an attempt to destroy these key bridges. After the first strike involving 79 warplanes dropped more than 100 750-pound bombs, minor damage was inflicted on the Thanh Hoa Bridge. In the days that followed, another 500 750-pound bombs were dropped, along with 30 Bullpup guided missiles. The result was five U.S. aircraft lost, and the bridge over the Chu was repaired and operating normally within days of the final attack.

Sturdy French construction defeated the tiny warheads on the Bullpup missiles. Speed, altitude, and anti-aircraft defenses thwarted the "iron bombs"—for it proved next to impossible to strike a steel span from a fighter or bomber racing by at more than 500 miles per hour while dodging anti-aircraft fire and enemy fighters. After many ineffective strikes with many losses, the limitation of aerial bombardment soon became apparent to even the most avid proponent of air power. Though the Vietnam War may have been won had those twin spans been destroyed, the combined might of the Air Force and carrier-based Navy fighters never succeeded in obliterating those bridges. Air attacks on these spans were halted after 1967.

On April 27, 1972, as U.S. involvement in the Vietnam War was winding down, there occurred an event that would change the future history of warfare. On that day, a brand-new cutting-edge weapon was first deployed in combat. Though its debut came with little fanfare, and far too late to alter the outcome of the war, the success of that weapon would have a profound effect on military thinking for the next 20 years, culminating in America's overwhelming and decisive victory in the Persian Gulf War.

On that fateful day in April, a dozen F-4 Phantoms flying out of the Eighth Tactical Fighter Wing in Thailand attacked Thanh Hoa Bridge for the first time in nearly six years. This time, the Phantoms carried two new types of guided "smart" bombs. One was an optically guided bomb with a television camera in its nose and vanes to control its descent. The second was a laser-guided bomb that required the illumination of the target by a low-power laser beam. This second type of smart bomb would follow the laser light right down to the target, adjusting speed and trim with vanes and fins. Both types of smart bombs had advantages and disadvantages. The first type was a fire-and-forget weapon, meaning that once it was fired, there was no way to change its target. Before it was released, the television camera had locked onto the target. After it was dropped, the vanes used gravity and inertia to guide it in. The laser-guided bomb relied on a second aircraft to paint the target and loiter in the area, continually bathing laser-light onto the target until the bomb impacted.

The weather turned out to be cloudy over the bridge, and the laser weapon wasn't used. But five optically guided bombs were dropped. All of them struck, and the railroad tracks were rendered unusable. Just over two weeks later, on May 13, the bridge was attacked with precision bombs again. This time a portion of the span crashed into the Chu River, and the rest was knocked off its foundation. The Thanh Hoa Bridge was out of operation after only *two* strikes, with no American losses. If anyone had doubts about the effectiveness of this first generation of precision-guided munitions, those doubts were erased after the attack on the Paul Doumer Bridge. On May 10, 16 F-4s armed with 20 laser-guided bombs and seven optically guided bombs successfully hit the Doumer span at least 16 times. A follow-up raid the next day dropped another eight laser-guided bombs. Again, there were no American losses, and the Paul Doumer Bridge was utterly destroyed. The spans were so damaged that it didn't carry rail traffic again until the United States withdrew entirely from the war, and the North Vietnamese—with Chinese and Russian help—rebuilt the span at great expense. The dawn of precision-guided munitions had arrived.

In Chapter 3, we saw the result of only two decades of research and development in precision-guided munitions on the outcome of the Persian Gulf War. Add 10 years of accelerated research and development on top of that and you begin to see the potential of intelligent weapons flexing its muscle as it enters technological adolescence. Yet despite the dramatic results of the first-generation smart bombs tested on the Vietnamese bridges, these weapons were only half of the equation—the warplanes that delivered these precision-guided weapons still had to survive the hostile skies long enough to do it. In the raids on the Thanh Hoa and Paul Doumer bridges, dozens of aircraft were required to accomplish the mission; yet, only a few of them actually dropped the bombs. The rest of the aircraft were decoys that dropped chaff to fool air defenses, Wild Weasels that followed radar waves to their source to obliterate anti-aircraft and surface-to-air missile sites, and AWACs (Airborne Warning and Air Control aircrafts) to coordinate the complex and multi-layered attack. Though Douhet's dream was destroying the target in a single raid (and a follow-up raid just to make sure), the air strikes were still costly and inefficient. Too many aircraft had to be placed in harm's way, all to place a few tons of munitions onto a target.

Precision-guided munitions *were* effective, but the aircraft that delivered them were dangerously close to obsolescence. Radar, surface-to-air missiles, enemy fighters, anti-aircraft fire, jamming, and the constant danger of malfunction made these aircraft an endangered species in the skies around Hanoi. If B-52 raids had not decimated much of North Vietnam's defensive capabilities prior to the bridge raids, the results may not have been so decisive. The Air Force was satisfied with the continuing evolution of smart munitions. But they had to find some way to make warplanes less vulnerable to the growing array of Soviet-designed countermeasures. By 1975, the military was close to discovering the other half of the most successful marriage in modern war—the union of precision with *stealth*.

The term *stealth* is applied to aircraft or missile systems that have been designed to produce as small a radar signature as possible, and also to reduce other "observable" characteristics—such as noise, heat signature, electromagnetic emissions, and naked-eye visuals—to a bare minimum. The term *stealth* can also be applied to ships and even land-based weapons systems, and has become a catch-all phrase used instead of the more precise military term *low observable*.

Camouflage paint schemes were the first attempt at creating low-observable aircraft during World War I, and dazzle paint schemes served a similar purpose on warships. But theoretical studies conducted during World War II also indicated that an aircraft's radar signature might be reduced through the use of special construction materials. Near the end of that war, Nazi scientists had developed a unique aircraft, the Gotha Go-229, a jet-powered flying wing—a peculiar aircraft configuration that has wings and a tail stabilizer, but no fuselage—made of radar-absorbent panels created from layers of charcoal, see-through fabric, sawdust-packing, and plywood. The war ended before the Gotha could be deployed, and it was only in the 1970s, after years of independent testing, that Lockheed designers realized how truly revolutionary the Go-229 really was. In the 1940s and 1950s, American designers mimicked the Go-229 to create their own prototype. Experiments with the Northrop YB-49 Flying Wing revealed that this basic shape gave the aircraft stealthy characteristics, despite the use of conventional all-metal construction and eight turbofan engines that spewed smoky exhaust.

In the late 1950s, the U.S. Air Force developed a series of radar test ranges where proposed airplane designs were set on poles or dangled from cables and then radiated by radar emitters to assess their *radar signature*—how the aircraft appeared on radar screens. As the radar waves bounced off the aircraft and returned to their receivers, aeronautics designers were able to determine what shapes resulted in a low radar signature, and what areas of an airplane produced the greatest radar-return ratio. This rate of return determines an aircraft's *radar cross-section* (RCS). A radar cross-section is the *apparent* size of an aircraft as it appears to search and fire-control radar systems. The RCS has nothing to do with the actual size of an airplane—a B-2 Stealth bomber is huge, yet on radar it presents an RCS that resembles a flock of birds or a soccer ball.

In World War II, the Germans discovered that using nontraditional materials in the construction of aircraft meant that more radar waves were *absorbed* than reflected. The British made a similar discovery inadvertently. When they deployed the Mosquito attack bomber, they learned that its all-wood construction resulted in a marked decrease of the Mosquito's RCS. But radar absorption by itself was not enough to ensure radar invisibility. Propellers, the edges of the wings and tail, the length of the fuselage, and the protruding canopy all caused a high rate of radar reflection, creating a large radar signature. To achieve true stealth, a less-reflective aircraft configuration would have to be found. But discovering radar-reflecting shapes that were also aerodynamic was no easy task.

By the 1970s, "pole testing" by the U.S. Air Force and Lockheed produced a breakthrough: They discovered that a slender, triangular-shaped delta configuration with a flat, faceted surface resulted a striking reduction in RCS. Lockheed's design team at the super-secret facility called the "Skunk Works" dubbed this singularly unaerodynamic design the Hopeless Diamond, and went to work designing an aircraft that could exploit this discovery.

Research to create a truly stealthy aircraft continued in the 1970s, both in America and in the Soviet Union. In fact, many key breakthroughs in stealth technology originated from the work of Pytor Ufimtsev, Chief Scientist at the Moscow Institute of Radio Engineering. Ufimtsev's research was dismissed by Soviet military planners, and his theories were actually translated and published in the West. Throughout the Cold War, the Soviet Union failed to exploit stealth technology. Meanwhile, by 1976, designers in America had grasped the three basic requirements for creating a stealth aircraft:

- The avoidance of design features that create strong reflections in the direction of the radar
- The use of materials (epoxy composites rather than metals) that absorb rather than reflect incoming radar energy
- Design features that mask or cancel out any remaining reflections (such as composite hoods to shield cavities, such as intake and exhaust ports and flat canopies, or a curved configuration without a fuselage or vertical stabilizer to reflect radar waves)

The first modern stealth aircraft was the Lockheed U-2 "Spy Plane" designed at Skunk Works. Working in total secrecy, the design team built the first prototype of the U-2 in just eight months. Its first flight came on August 1, 1955. Less than a year later, on July 4, 1956, the U-2 made its first reconnaissance flight over the Soviet Union. It was soon apparent that even though the Soviets could not shoot it down because of its speed and high operational altitude, the Russians knew the U-2 was up there because they could track it on radar.

SNIFFING OUT SKUNK WORKS

Based in Burbank, California, Lockheed's Skunk Works is a now-legendary project office where small, highly motivated design teams worked with a minimum of organizational constraints and even less bureaucracy. Off limits to most employees of Lockheed, and isolated from their main facilities, the Skunk Works has been located in a seedy industrial plant since its inception in 1943. The brainchild of aeronautical designer Clarence "Kelly" Johnson, the corporate ethos was banished at the Skunk Works—people wore casual clothing, and the organizational hierarchy was equally relaxed, even though Kelly Johnson was considered a harsh taskmaster. In essence, the Skunk Works represented the ultimate vision of the "boffin"—the civilian scientist that soldiers fighting modern technological wars rely upon to give them the edge.

Finally, after several modifications, including the application of radar-absorbing black paint, the Soviets shot down a U-2 piloted by

Francis Gary Powers on May 1, 1960, using a new, advanced surface-to-air missile. This downing caused an international incident that threatened the credibility of the Eisenhower administration, which had previously denied flying spy planes in Soviet airspace. A better aircraft was soon ordered by President Dwight D. Eisenhower, who rejected supersonic speed and high altitude as the only means of achieving stealth.

The next aircraft produced by the Skunk Works was the SR-71 Blackbird, which remains the fastest manned aircraft ever to fly. The SR-71 was an example of an aerodynamic design that was rendered moderately stealthy by the addition of a coating of radar-absorbing material. The Blackbird had a low radar signature at various angles and frequencies of the radar spectrum—especially the X-band, where Soviet surface-to-air fire-control radar operated. Though the SR-71 Blackbird served as a high-level reconnaissance aircraft for nearly 35 years, it was never a true stealth airplane. A scientific breakthrough was needed before one could be developed.

That breakthrough came in 1975, when the Air Force Foreign Technology Division translated a paper by Pytor Ufimtsev called "Method of Edge Waves in the Physical Theory of Diffraction." This study was read by Skunk Works mathematician Denys Oberholser and was handed to the new director of the design division, Ben R. Rich. "[Oberholser] presented me with the Rosetta Stone breakthrough for stealth technology," Rich said later.[1]

Denys Oberholser realized from reading Ufmitsev's study that if an aircraft could be built with a surface made up of flat, triangular shapes, its RCS could be significantly—perhaps dramatically—reduced. In conjunction with a Skunk Works designer named Richard Scherrer, Oberholser devised a design shaped like an arrowhead that had a radar signature of less than a thousandth of a conventional aircraft. Scherrer and Oberholser called it the Hopeless Diamond.

This breakthrough came at an opportune time for Lockheed. The U.S. Defense Advanced Research Projects Agency (DARPA) was interested in stealth technology, and had given several smaller companies millions of dollars to develop such technology. Though Lockheed was not one of those favored corporations, they were permitted to join the competition to design and build the first true stealth aircraft. Tests at White Sands, New Mexico, demonstrated that the Hopeless Diamond configuration returned an RCS smaller than the size of a golf ball, and by April, 1976, Lockheed was declared the winner. The Skunk Works was handed $30 million to design and build two XSTs (Experimental Stealth Tactical, or Experimental Survivable Testbed) prototypes named "Have Blue."

From this humble beginning came the most successful black program in military history. Black programs—more formally known as "Special Access Programs" (SAPs)—are top-secret military projects that have tight security and rigorous restrictions on personnel with access to information on the project, its funding, and its goals. High-technology projects like the F-117A and B-2 stealth aircraft, or the Strategic Defense Initiative, are most often singled out for "special access." In addition to secrecy, misinformation is employed to keep a black program secret.

In November 1978, a secret contract was awarded to Lockheed to build the first batch of five stealth aircraft for the U.S. Air Force. That number was soon expanded to 25 aircraft, and General Electric was awarded a clandestine contract to develop a non-after-burning version of their popular F404 engine to power the new super-secret aircraft. The code name for these projects was Senior Trend. The first F-117 Stealth fighter was completed in May 1981 and flew a month later. The initial batch of five were used for aerodynamic and propulsion tests. The first production example of the F-117 crashed on its 1982 maiden flight because of an error in its construction, and not because of any design flaw. The first F-117A was delivered to the Air Force in August 1982.

Meanwhile, the first squadron to fly the stealth plane—the 4450th Tactical Group at Tonopah, Nevada—had already been established back in 1979, and by 1984, there were four squadrons flying F-117s, now designated the "Nighthawk." Security on the bases around Las Vegas was tight—the airplanes flew only at night, and were kept inside hangars with locked doors until after sunset.

Soon, the first cracks in the wall of secrecy surrounding Senior Trend began to appear in 1984, when the professional journal *International*

Defense Review published an article titled "Stealth" by veteran aviation journalist Bill Sweetman. By 1986, classes at Georgia Tech and the American Institute of Aeronautics on the subject of stealth technology were the basis for two textbooks, further disseminating formerly classified research.

TOYING WITH STEALTH TECHNOLOGY

The biggest shock to the guardians of the F-117s stealth technology came in July 1986, when the Testor Corporation of Rockford, Illinois, marketed a $9.95 molded black plastic hobby construction kit that claimed to depict the "Lockheed F-19 Stealth Fighter" in ½ scale. *The Washington Post* and *The New York Times* both featured news stories about the model kit that hinted national security had been breached. Following this basic design, Hasbro issued a similar-looking "GI Joe Stealth Plane" in the popular toy line a few months later.

But what appeared to be a horrendous security leak was most likely a classic case of misinformation, because both the model kit and the GI Joe aircraft featured a singularly unstealthy rounded "Frisbee" design with a long conventional fuselage that was totally at odds with the faceted delta configuration Lockheed was actually producing. The only similarity between the F-19 kit and the real F-117 was that they were both tinted black. As with everything concerning stealth technology, appearances were deceiving.

Finally, in late 1988, the F-117A Nighthawk was unveiled to the public in a highly publicized rollout. One reason for lifting the veil of secrecy was to draw attention away from another super-secret stealth warplane project being developed by Northrop—a project that remained classified throughout the rest of the 1980s.

The F-117A Nighthawk was the direct result of the successful XST tests by Lockheed's Skunk Works. The Nighthawk is a single-seat strike fighter with a length of 66 feet and a wingspan of 43 feet, 4 inches. It can haul 4,000 pounds of munitions—all carried internally to maintain the aircraft's stealthy characteristics—and reaches a subsonic top speed of Mach 0.9. To keep the radar cross-section at a minimum, the entire fuselage is faceted with straight lines—not curves—dominating the warplane's arrowhead configuration.

Faceted panels of classified composite material are mounted on a skeletal frame, and then the whole is covered with radar-absorbing material. Even the airfoil is faceted to diminish RCS, and the canopy and removable panels are also faceted with a saw-tooth design to further decrease the radar signature. The cockpit is located inside the fuselage to eliminate the need for a bubble canopy that would reflect radar. Engine intakes have screens to block radar energy from entering the scoop. Incoming air is mixed with the exhaust to significantly reduce the aircraft's heat signature.

The basic "Hopeless Diamond" configuration and the flat, faceted panels severely diminish the aerodynamic qualities of the Nighthawk, so computers and a GEC Astronautics quadruple fly-by-wire control system are required to make the aircraft flyable. For navigation, the Nighthawk carries a Texas Instruments Forward-Looking, Infrared (FLIR) sensor mounted in a screened cavity just below the flat-paned front canopy, and a Downward-Looking, Infrared (DLIR) and laser designator for painting a target are located in a screened cavity in the belly.

Holes, gaps in joints, and especially cavities like intake ports and weapons' bays, cause a larger radar signature because energy gets inside and bounces around before radiating back to the receiver. The cavities on the Nighthawk had to be screened, but the screens first proposed caused major headaches for the designers. Originally, transparent covers for the FLIR and DLIR (and later IRADS) cavities were to be made of a zinc-selenide crystal material that had to be grown over a six-month period at a cost of $300,000 a unit. When this design component proved impractical, the transparent window was replaced with a very fine wire mesh used on the intake ports. To mask its signature, the canopy transparencies were coated with a layer of indium-tin oxide (ITO) that blocked infrared radiation.

The Nighthawk has a unique butterfly tail that eliminates the need for a large, projecting radar-reflective vertical stabilizer. However, the lack of a vertical fin makes the stealth fighter unstable at high speeds.

In fact, the Nighthawk's stealthy design created an aerodynamic nightmare for its inventors. The F-117A is slow and not very maneuverable when compared to top-of-the-line supersonic fighters, and it would be quite vulnerable if caught by an enemy fighter in the sky. That is why the Nighthawk has an all-black scheme and primarily operates in darkness—it is undetectable to radar *and* to the naked eye.

The F-117A is not a fighter, so its name is actually a misnomer. The Nighthawk is an attack plane, designed to deliver precision-guided munitions to a target deep within enemy territory while avoiding detection.

Though the Nighthawk is barely two decades old, it has had three upgrades—all to internal hard and software systems. The most recent modification replaced the old FLIR and DLIR sensors with the brand-new Texas Instruments Infrared Acquisition and Designation Sensor (IRADS), mounted on an internal turret below the cockpit, and its obsolete navigation system with the Ring Laser Gyro Navigation Improvement Program (RNIP) that included a Global Positioning System (GPS) satellite receiver.

Though the Nighthawk performed very effectively during its first combat action over Panama in Operation Just Cause, and quite spectacularly in the skies over Baghdad in the Persian Gulf War of 1990 and 1991, the F-117A must be considered a first-generation—and therefore highly imperfect—stealth aircraft. The next generation of stealthy warplanes, such advanced designs like the F-22A Raptor and the troubled Joint Strike Fighter (JSF) will possess increased speed, maneuverability, range, and striking power, and will also serve as a true fighter capable of engaging other high-performance warplanes in the hostile skies of the 21st century. In reality, rapid developments in stealth design and other recent innovations rendered the F-117A Nighthawk nearly obsolete by the time it saw combat in Desert Storm.

In 1981, as President Ronald Reagan began a massive buildup in American forces to counter the growing threat of the Soviet Union, Lockheed was given its contract to build what would become the F-117A Nighthawk, and another contract to develop cruise missiles with stealthy characteristics. Additionally, Congress directed that a new strategic bomber be designed to replace the B-1B program hastily cancelled by the Carter administration, a cancellation that left a strategic gap in America's long-range bomber capabilities.

This time it was Northrop who won the contract, worth more than $70 billion, primarily because—like Lockheed—they had been working secretly to develop nonobservable technology throughout the late 1970s. Northrop proposed a design that was already known for its stealthy characteristics: the flying wing. Northrop had built its first flying wing design in 1947, but the aircraft proved unstable and unable to pull out of a spin. In testing its maneuverability and stall and spin limitations, two test pilots perished in a crash. Recent innovations like computer-assisted controls and fly-by-wire digital systems resuscitated the tricky wing configuration and made it viable once again.

Boeing, the company that built the B-17 Flying Fortress during World War II, had also looked at tailed and tailless delta-shaped con-figurations in the 1970s in an attempt to exploit the fact that radar energy will diffract off flat, horizontal surfaces. Teaming up with Boeing, Northrop developed an aircraft design optimized for maximum range and payload, a bomber that could penetrate deep into Soviet airspace to launch nuclear-tipped cruise missiles at strategic targets. Though much of its history is still classified, it is believed that the first prototype of the stealth bomber flew in late 1982. Design problems—primarily with the cockpit transparencies and the honeycomb composites used in its construction—delayed the rollout of the first Northrop B-2 Advanced Technology Bomber—dubbed "Spirit"—until July 17, 1989. By then, the Soviet military threat that the Spirit had been designed to counter had virtually ceased to exist.

The Northrop B-2 Spirit is the first strategic bomber acquired by the U.S. Air Force since the B-1B in the mid-1970s. The warplane was originally meant to be high-altitude bomber, and significant design changes were required to make it a low-level capable strike aircraft as well.

ONE B-2 AT A TIME

Work on each B-2 begins at a converted Ford Motor Company automobile assembly plant in Pico Rivera, California, where nearly 7,000 employees work on each unit. Final assembly is done at a brand-new Northrop facility at Palmdale Airport, California, in gigantic screened hangars.

In a sense, each aircraft is a prototype, for little is done on an assembly line. Most of the work is done by hand and is quite labor intensive. The honeycomb composite materials that make up the aircraft's skin is difficult to pierce—power drills tend to make oblong holes that create gaps that diminish the B-2's stealthy characteristics. Designers and assemblers have had to design and fabricate some parts on the spot, further adding to the aircraft's expense, which exceeds half a billion dollars per unit.

When the first B-2s entered service, gaps in joints and around access doors were caulked or covered with a special tape, but in flight the tape tended to peel away and cause radar reflection. So the aircraft is given a coating of improved anti-radar coating made from radar-absorbing materials called MagRAM, an elastomer (elastic polymer) filled with iron, which stores radar energy rather than conducts it.

The first Spirit built was rolled out in November 1988 and made its maiden flight in July 1989. Six "developmental test prototypes" were built, and all of them were later placed into operational service. The Northrop B-2 Spirit is 60 feet long, with a wingspan of 172 feet. Powered by four GE F118-GE 100 turbofan engines, the Spirit can cruise at a high-subsonic speed of approximately Mach 0.89, at a ceiling of 50,000 feet, at a range of more than 6,300 nautical miles, while delivering a payload of 40,000 pounds of ordnance stored in internal bays.

The aircraft has a simple front profile featuring two moderately swept leading edges that meet at the low, rounded nose. The trailing edge, in a saw-tooth configuration that features 14 straight surfaces aligned in one of two flight angles, ensures that radar energy is reflected away from the rear of the aircraft. There is no vertical stabilizer, which greatly increases its stealthy characteristics. Instead, each wing has a drag rudder and elevon on the outboard trailing edge, and two more elevons on the next inboard section, while the central beaver tail forms another moving surface to control the aircraft's flight. Needless to say, this revolutionary design makes the aircraft aerodynamically unstable, so a General Electric quadruplex digital flight control system that incorporates fly-by-wire controls—plus a stability-augmentation system—have been incorporated into the B-2's design. Despite the unique configuration, the B-2 is highly fuel efficient, and requires far fewer tankers for airborne refueling than its predecessors, the massive B-52 and the sleek B-1B.

The engines are buried deep in the fuselage to diminish noise and heat. Inlets on the upper wing take in air for the power plants, while a secondary inlet mounted in front of the main one draws in cool air to mix with exhaust to reduce the rear heat signature. The exhaust is also spread laterally, through a network of recessed ports in the upper wing surface, to further reduce the Spirit's infrared signature. Originally designed to contain a contrail suppressor, the design was changed to include a rearward-facing radar contrail detector, which warns the pilot when he is leaving a trail in the sky so that he can alter altitude to elude detection. The cockpit configuration and avionics package are still highly classified, but probably includes a GPS system; FLIR, DLIR, or IRADS; a digital-targeting system; the previously mentioned quadruplex fly-by-wire systems; and other highly advanced controls employing flat multifunctional displays for use by the three-man crew.

The Air Force was originally supposed to procure 133 B-2s, but the end of the Cold War caused a significant reduction in that number, and only 21 have been produced. A follow-on batch of 20 more B-2s was vetoed by President Bill Clinton in the mid-1990s, and production was halted. The 509th Bomber Wing was established at Whiteman Air Force Base in Missouri on April 1, 1993, to operate the B-2. The first operational unit to receive the aircraft was the 393rd Bomber Squadron. In 1998, the B-2 was deployed to Guam and flew its first combat sorties over Kosovo in 1999.

The B-2 was built to be sturdy and durable and has an operational life span of more than 40,000 hours. If large numbers of them had been built, the aircraft could have conceivably outlasted the 50-year-old

B-52 fleet. But because so few B-2s were actually constructed, the shelf-life of the Spirit will probably be determined by the attrition rate of the aircraft in service. In 2001, the new George Bush administration tentatively alerted Northrop Grumman that it would like to procure 40 more Spirits for long-range deployment. But until—and if—that happens, with only 21 operational B-2s, the Air Force plans to keep the aging B-52H in service until 2040, giving this stalwart design an 80-plus-year operational life span unrivaled in the history of military aviation.

The first high-performance, air-superiority stealth fighter to approach operational service is the Lockheed Martin F-22A Raptor. This warplane represents a quantum leap in stealth aircraft design—a design that is both stealthy and fast, and is capable of holding its own against any competitor in the sky. If the Raptor enters production as scheduled and becomes operational in 2005 and 2006, it will become America's front-line fighter for the 21st century, replacing the stalwart F-15 Eagle. The Raptor will ensure U.S. Air Force dominance of the skies for the next 50 years and will defeat threats that the aging F-15 will no longer be able to counter.

After beating out its nearest competitor, the Northrop/McDonnell Douglas YF-23, the F-22 Raptor won an Air Force development contract on April 23, 1991. Although the overall shape of the F-22 is fairly conventional, many design features were dictated by the need for stealth. Reduction in the Raptor's radar cross-section is achieved by the sloping sides of the fuselage and canopy and canted vertical tail surfaces. Platform alignment—meaning that the leading and trailing edges of the wing and tail have identical sweep angles—helps to reflect radar energy into only a few directions, further reducing RCS. Like the F-117, the edges of the canopy, weapons bay doors, and access panels are saw toothed in design to further reflect radar energy. A twisted intake duct conceals the front face of the engines from radar while the avionics antennas are located on the leading or trailing edges of the wing and fins. Many of these externals are flush with the surfaces.

The Raptor is 62 feet long, with a wingspan of 44.5 feet. Weighing in at more than 16 tons, the Raptor can achieve a flight ceiling of 66,000 feet and a range of nearly 2,000 miles. It is powered by two Pratt and Whitney F119-PW-100 turbofan engines, each providing 39,000 pounds of thrust. Like most modern warplanes in America's arsenal, the Raptor is a team effort because it is too expensive and risky for one corporation to go it alone in the world of high-tech weapons procurement. Lockheed Martin Aeronautics Systems is responsible for the overall weapons systems integration. Lockheed Martin Tactical Aircraft Systems is responsible for the mid-fuselage, for armaments, electronic warfare systems, and the integrated communications, navigation, and identification system (CNI). Boeing is responsible for the wings and aft fuselage, plus internal structures that hold the engines and nozzles and the radar system.

(Photo courtesy Lockheed Martin)

The F-22 Raptor.

The Raptor is controlled by a triplex, digital, fly-by-wire system. Besides being stealthy, the F-22A has thrust vectoring—nozzles that adjust in flight to provide the kind of super-maneuverability found on the most-advanced Russian and European designs such as the Sukhoi Su-27. Two-dimensional convergent/divergent engine exhaust nozzles capable of moving 20 degrees upward or downward in the vertical plane are used to augment aerodynamic pitch control, particularly at reduced speeds and at high angles of attack. Thrust vectoring allows jet aircraft to do amazing things. The Sukhoi Su-27 can literally stop in the air and dance on its tail, and, for a few short seconds, it can even fly backward.

The F-22 is unmatched by any other fighter aircraft in service or in development because it combines stealth, integrated avionics, and super-cruise characteristics. Supercruise, or supersonic cruise, means that the F-22 is capable of cruising at least 50 percent faster than any front-line fighter in operation today—without using full military power. The maximum speed is reported to be between Mach 2 and Mach 1.9, which is slower than the F-16 Fighting Falcon and the F-18 Hornet, but this does not take into account the F-22's supercruise capability. Using supercruise technology, the Raptor can be flown at three to six times the supersonic endurance of any other fighter, even without augmentors like an afterburner. Conventional fighters like the F-16 can attain great bursts of speed, but cannot maintain those speeds for more than a few minutes. Supercruise allows the F-22 to maintain a high rate of speed for a long time, meaning that an F-22A can maintain Mach 1.5 for more than 30 minutes—longer than the aircraft is likely to be over a target area or in hostile airspace, making it far less vulnerable than conventional fighter and attack aircraft. Though its exact capabilities are still a guarded secret, it is known that the Raptor can match or exceed the agility of any front-line fighter in existence.

The configuration of the F-22 is dictated by the demands of stealth, supersonic cruise, and agility. The clipped-delta wing is efficient at high speed and is lightweight, even though the Raptor is larger than the F-15, with a broad hull and massive internal fuel tanks. Unlike other stealth aircraft, only about a quarter of the F-22's airframe is made of composite materials—most of the rest is comprised of titanium. The inherent maneuverability allows the Raptor to achieve extreme angles of attack—even straight up or straight down—while remaining under full control.

Inside the cockpit, the Raptor is also highly advanced. The pilot is provided with a sensor fusion system that combines data from many different sensors on a single display system so that he does not have to rely on many different displays to get a clear picture of the battle environment. That single unit is the 8-inch by 8-inch Tactical Situation Display (TSD), which includes three more 6-inch display screens left (for defense), right (for attack), and below, the center monitor displays flight control statistics. In conventional warplanes, the radar and electronic warfare systems, navigation, communications, identification, and targeting systems are separate, but the F-22A's sensors buck that trend. Although they operate independently, these systems are integrated into the Raytheon Common Integrated Processor (CIP), which consists of two banks of liquid-cooled computer modules located in the forward fuselage.

All critical information is supplied by the CIP through the Tactical Situation Display. The Raptor also contains two high-performance sensors: the Northrop Grumman APG-77 radar and passive electronic surveillance measures (ESMs) built into a BAe Systems ALR-94 electronic warfare system. The ALR-94 provides 360-degree coverage in all radar bands, and any target trying to use its radar to search for a Raptor will be detected, tracked, and identified by this advanced system at a range of over 250 miles—long before the F-22A is itself detected.

As targets are detected—either by the F-22A's internal radar or from high-flying AWACs, the Raptor's advanced software assigns them to a "tracking file," and as other on-board sensors pick up more information on the target this will also be added to the tracking file until a clear picture of the threat is presented to the pilot. The Raptor's proposed armaments consist of a mix of existing and planned medium and short-range air-to-air missiles and a 20mm M61A2 internal cannon made of composite materials, with a recessed barrel instead of one that protrudes, to ensure stealth. AIM9 Sidewinder missiles will be carried in the internal bay located in the external sides of the intake ducts. "Other weapons" (still classified) will be carried in the ventral weapons bay. Four underwing storage areas can carry 5,000 pounds of additional ordnance, but until stealth munitions are developed, these stores will render the Raptor far less stealthy if used. The F-22A has a ground attack capability and can launch and direct precision-guided ground attack weapons such as the GBU-32 Joint Attack Munition (JDAM).

The Air Force originally planned to acquire more than 600 F-22As, but budget cuts have lowered procurement numbers. The Air Force claims it will be content with about 339 Raptors, manufactured at a rate of 36 aircraft per year. Any further reduction in these numbers may result in a cancellation of the entire program. Operational locations for F-22A Raptor squadrons have (as of this writing) not been announced.

Lockheed Martin has created an aircraft that is conservative in appearance, but remarkable in its capabilities—an aircraft that is stealthy, agile, and ferocious. When the first Raptors reach operational status in 2005, this aircraft will represent "the greatest single advance in fighter design since the advent of the jet."[2]

For many years, an argument has raged over the range of strike capabilities of military aircraft. Some in the defense community believe in economy, and maintain that bombers and ground attack aircraft should have the ability to defend themselves against enemy fighters and have the potential to fly dangerous missions deep into enemy territory without fighter escort. Others maintain that attack plans and bombers should only perform ground attack missions, and that a separate weapons platform—a squadron of fighters or fighter interceptors—should be delegated the task of defending the bombers.

For the Navy, there is no option. Carrier-based aircraft must serve as both fighters and ground-attack planes because they are usually the only airborne resources in that particular theater. And in the past, the Air Force has also demanded that their front-line warplanes be dual-mission capable, despite the fact that large-scale air campaigns like the Gulf War and the air war over Kosovo have involved the use of task-specific aircraft, controlled and directed by tacticians manning AWACs aircraft, and have a bird's-eye view of the battle. Even the F-22 Raptor is designed to fulfill a ground-attack role, because the aircraft it will eventually replace—the F-15 Eagle—is also a dual-purpose aircraft.

But the demands of the modern battlefield have increased the need for specialized aircraft. Today's missions require bombers and attack planes to hit the target; fighters to guard them; Wild Weasel aircraft to seek out and destroy radar, anti-aircraft, and surface-to-air missile sites; radar and radio jamming aircraft to interfere with enemy command and control; airborne tankers to refuel the warplanes before the strike; AWACs to orchestrate the battle; and—if precision-guided munitions are used—aircraft with laser-designation pods to paint the target for the follow-on missile attack.

Now, this is an awful lot of aircraft, crews, fuel, and effort to place a few tons of munitions onto a target. The use of precision-guided bombs has helped somewhat. During World War II, 9,000 bombs were required to hit a given point (true saturation bombing). In Vietnam, jets could strike more accurately and at lower levels than B-17s, B-24s, and B-29s, and that rate was diminished to a mere 300 bombs. During Desert Storm, when precision-guided munitions made their debut, two or three bombs were sufficient to level a target.

While these numbers represent a tremendous breakthrough, it must be remembered that the weapons deployed in the Gulf War were only the second generation of precision bombs. More and better precision-guided munitions are being designed and are ready for deployment right now. The precision characteristics of this class of weapons are built into them. Once launched, PGMs can fly faster and truer than manned aircraft, and will strike their target far more accurately than bombs guided by gravity. In fact, during an air attack, the attack planes are far more vulnerable than the precision-guided missiles and bombs they deploy because the PGMs, once launched, are very difficult to stop.

This raises a fundamental question: Do precision-guided munitions require manned aircraft for delivery?

If a human being is to remain the center of gravity in an air battle, he will need an effective weapons platform to do it. Decades of battle-field evolution have worked to increase the survivability of the manned warplane—from active defensive capabilities like machine guns and air-to-air missiles, to passive technologies like camouflage and stealth. Speed and maneuverability have also been decisive factors effecting survivability, but with the addition of digital, fly-by-wire controls, thrust-vectoring, scram-jet technology, and advanced aerodynamic designs like variable-swept and forward-swept wings, designers have reached the limit of speed and maneuverability in manned aircraft. This limit has not been imposed by technology but by the limitations of the pilot and the human body's ability to endure the stresses of high-speed air combat.

In the past, pilots had to be careful not to maneuver too suddenly or risk putting excessive pressure on the aircraft's structure. An abrupt turn or dive could rip the wings from a World War II or Korean War–vintage aircraft. Vietnam-era jets were better, but violent maneuvers could still result in disaster. This is not true of the current generation of front-line warplanes. The F-16 Falcon and F-18 Hornet can withstand rapid accelerations and sudden turns that the pilot cannot, for the human operator is susceptible to a condition known as gravitational loss of consciousness, or G-LOC.

The human body is not built to suddenly accelerate to eight or nine times the force of gravity at sea level and still function properly. At nine g's, a 175-pound man can suddenly be hit by more than 1,400 pounds of pressure in a split-second. This can cause a sudden decrease of the blood pressure to the head, which results in a loss of consciousness or a blackout. Even if a blackout does not occur, sudden g-force can cause narrowed or spotty vision, difficulty in breathing or movement, and temporary disorientation. G-LOC strikes without warning, and can be fatal in a high-performance jet aircraft, for in combat conditions the pilot must battle G-LOC even as he seeks to elude air-to-air or air-to-ground missiles and enemy fighters.

Precautions can be taken. For example, pilots wear pressure suits that inflate to constrict the lower extremities, forcing blood to the head and heart. And the pilots themselves must take precautions. They are required to be in excellent physical condition, and they must exercise constantly and sometimes eat a special diet to lower their blood pressure. There are even revolutionary new cockpit configurations being considered, designs that will feature flight suits filled with water to absorb g-force. Precautions have also been built into contemporary high-performance aircraft, including monitors that take over control of an aircraft if they sense that the pilot has experienced G-LOC or is wounded in battle.

Yet, even these extreme measures cannot protect the pilot or the aircraft adequately in the hostile environment of modern war—not when an unmanned aircraft unencumbered by a human operator can perform 100 times more violent maneuvers to elude attack, or pull 10 times as many g's to avoid destruction by a missile. These characteristics alone seem to indicate that unmanned aircraft have the potential to be more survivable than manned aircraft are. Only one element is missing from this equation: the guiding intelligence. Right now, designers are racing to fill that gap.

Intelligent munitions are different from precision munitions. Precision-guided weapons—PGMs—still require a human inside the combat zone to launch the weapon from a manned aircraft, to paint the target, or to guide it in. Intelligent weapons such as the Tomahawk cruise missile require a human programmer to feed it targeting data, but once it is programmed and launched—usually from up to 800 miles away—a Tomahawk requires no more human intervention to successfully accomplish its mission, which it did more than 90 percent of the time in the Gulf War. The built-in computer and GPS system intelligently guide the Tomahawk to its destination without putting a human being in harm's way.

The Tomahawk has proved its worth in the Persian Gulf War (see Chapter 3), and in recent years a new generation of intelligent aircraft—not all of them guided bombs—have made their debut. In fact, the drone, or unmanned aircraft, may become a major weapon in the future of war. As a reconnaissance resource, the unmanned aerial vehicle (UAV) has become invaluable in the campaign to defeat the Taliban in Afghanistan after the September 11, 2001, World Trade Center attack. Carrying sophisticated cameras, radar, and active and passive visual aids like FLIR and DLIR, and in some cases armed with missiles, these semi-intelligent drones are controlled by operators hundreds of miles away and at a high altitude out of anti-aircraft missile range, and they can also work independently, using their internal programming.

The Predator is the most common type of drone aircraft in use today. Flying slow and low, the Predator carries a radar system able to detect objects as small as 4 inches across from up to 15 miles away. A smaller drone called the Gnat carries a high-resolution radar system and can remain aloft for up to 48 hours. The Predator and the Gnat were first deployed in Afghanistan, and the Predator was the first unmanned aircraft to launch an attack against an enemy concentration on the ground using air-to-ground Hellfire missiles.

UNMANNED IS UNMANLY

There are many debates swirling about the F-22, unofficially christened the Lightning 2, after the famed twin-engine P-38 Lightning of World War II. Some insist the F-22 is far too expensive. Others maintain it is a design for another age—specifically the Cold War era. Still others believe that the age of the manned aircraft will soon come to an end.

The computer intelligence guiding the Tomahawk cruise missile is stupid compared to the new generation of automated guidance systems currently in development. There are strategists who believe it is wasteful to risk a human pilot where a robotic aircraft can go. Of course, the U.S. Air Force has resisted such a change because in time such a transformation from manned to unmanned aircraft will render the aviator obsolete.

Traditions die hard, and none die harder than the legend of the hot-shot fighter pilot who spurns discipline, flies by the seat of his pants, and gets the job done in spite of enemy action and the vagaries of his superiors. Like the rogue cop, the misunderstood outlaw, and the lonesome cowboy, the pilot hero holds a special place in the annals of war, and in American popular culture.

But perhaps the most sophisticated intelligence drone presently in use is the Northrop Grumman Global Hawk, which can cruise around at about 60,000 feet for up to 24 hours. This unmanned bird contains still-classified avionics, surveillance, and radar packages, and a de-icing system for its wings. Research is currently underway to create stealthy drones that can perform the same functions as the Predator and Global Hawk, but which will be much more difficult to see or shoot down.

The first of these stealthy designs to reach the prototype stage is the X-45, which has been in development for years at a cost of around $256 million. The X-45 is an unmanned combat aerial vehicle (UCAV) with a capacity to carry more than 3,000 pounds of missiles or bombs to a target. Right now the Air Force is looking at the X-45 as a potential countermeasure against enemy ground-to-air missiles and radar installations, but the potential is there for a more wide-ranging set of combat tasks. The first X-45 is expected to enter service by 2010.

The X-45 is semi-intelligent and designed to act somewhat autonomously. Its "pilot" will be far from the scene and may be "flying" several drones at once. The Y-shaped X-45 has a large air-intake instead of a canopy, and from the side it has a slim, stealthy profile. The target cost of each unit is about $15 million, or about a third of the cost of a next-generation manned fighter plane like the F-22. The first two prototypes of the X-45 have been developed by the Defense Advanced Research Project (DARPA), by the Air Force, and by Boeing, who is the primary contractor.

According to an Associated Press article published in June 2002, Air Force researchers at Eglin Air Force Base in Florida have spent a year examining drone weapons platforms that may eventually replace the Vietnam-era AC-130 Spectre gunship, used primarily to support ground troops.[3] Initial tests have been so promising that some Air Force officials have recognized the potential for these drones in taking on missions performed by bombers and strike aircraft, too.

General John Jumper, Air Force chief of staff, created Task Force Warlord—a new, top secret experimental program to test the limits of unmanned aircraft—in 2001, after getting a preview of the potential of such aircraft while watching live video beamed to the Pentagon from Predator surveillance drones flying in the skies over Afghanistan. As General Jumper watched the Central Intelligence Agency's UAVs seek out and attack targets with Hellfire missiles in the first combat test for armed Predators, he was duly impressed. In Afghanistan, video links from the tiny Predator drones were hooked up with screens on AC-130 gunships, increasing the older aircraft's patrol and reconnaissance range—and firepower—dramatically through the use of these remote and remote-controlled "eyes in the sky."

Arming Predators and linking them to gunships were just two ideas hatched in the UAV Battlelab, a research facility at Eglin AFB, and the concept went from demonstration to combat within a year. It is hoped that the secret work of Task Force Warlord could become a reality just as quickly. Concepts proposed for future unmanned aircraft include better computers and sensors, and composite or ceramic parts to reduce weight and increase stealth, as well as artillery shells guided by satellites or by gunships, and more advanced data links and UAV control from the gunships.

In the near future, Air Force strategists envision swarms of armed drones launched from a "mother ship" attacking ground targets while other unmanned aircraft paint the targets with advanced radar that can see through rain, sleet, snow, smoke, and even foliage. Meanwhile, "sleeper weapons" land in remote areas deep behind enemy lines, to emerge and strike quickly at the enemy forces from close range, on their vulnerable flanks or even at their rear—disorienting command and control and inflicting damage on critical supply lines, vehicles, and troops. As computers get more sophisticated, such attack systems might be capable of independent action—making the era of the intelligent flying killer robot a reality. The are even various proposals with the Air Force to use "obsolete" airframes like the F-15 or F-16, retrofitted with intelligent guidance systems and reworked as unmanned drones. Armed with missiles or bombs, up to eight of these drones can be controlled by a single operator sitting in a remote location inside an AWAC, an F-22, in a command and control center on the ground, or at the Pentagon a thousand miles away. From this remote location, the operator can send the drones into harm's way to deliver their payload, without ever risking a pilot's life.

The reality of aerial warfare in the 21st century is a battle between supersonic, high-agility missiles going up against lower-speed, much-less-maneuverable manned aircraft. Modern combat aircraft may be capable of airspeeds in excess of Mach 2, but the missiles that kill them are smaller, more maneuverable, less observable, and can move at six times the speed of sound and sustain a g-force in excess of 100 times their own weight. It is not beyond the realm of possibility that future unmanned weapons will have these same capabilities—and will also have a range in the thousands of miles, and internal computer intelligence systems to guide them to their target. In time, such weapons may render the manned combat aircraft—like the battleship, and soon the tank and the aircraft carrier—totally obsolete.

CHAPTER 10

THE HIGHEST FRONTIER: WAR IN SPACE

The Global Positioning System (GPS) is a worldwide satellite navigational tool formed by 24 satellites orbiting the planet and their corresponding receivers on Earth. GPS satellites continually transmit digital radio signals containing data on the satellites' location and the exact time it takes for a signal from the satellites to reach its earthbound receivers. Based on this information, through the use of a GPS receiver, one can track a distant object or objects—calculating longitude, latitude, and altitude coordinates with pinpoint accuracy. This system is used to monitor weather patterns, wildlife habitation and migration, and the movement of people and things, including vehicles and aircraft. During the Persian Gulf War, this GPS system was also critical to the navigational functions of most U.S. weapons systems, from ships to tanks to cruise missiles to smart bombs to high-performance jet fighters—right down to the infantry squad.

During the Gulf War, virtually all communications in the theater of operations were accomplished through three orbital satellites belonging to the U.S. Defense Satellite Communications System (DSCS). One satellite was in geo-stationary (occupying the same position over the earth at all times) orbit over the Indian Ocean, a second had to be adjusted to access the Gulf region, and a third, reserve satellite was activated in case one of the others malfunctioned.

Military, governmental, and civilian weather satellites accurately followed climactic patterns so that meteorologists were able to provide the Navy and Air Force with fairly precise forecasts—making advanced planning of the air and ground war possible. Meanwhile, a highly classified, comprehensive system of orbiting reconnaissance satellites carried out a wide range of intelligence functions during the Gulf War, from tracking enemy movements to intercepting Iraqi communications to detecting the launch-plume from SCUD missiles.

Though cruise missiles, stealth aircraft, smart bombs, and General H. Norman Schwarzkopf's "Hail Mary" flanking maneuver made headlines, none of these weapons or tactics could have been employed if the Iraqis had somehow found a way to knock our satellites out of orbit, or interfered somehow with the vast array of data being gathered outside of Earth's atmosphere. The complex system of intelligent, computerized objects with eagle-eyes, multi-spectrum viewing, and ears like a hawk were crucial to monitor the activities of friendly and hostile forces alike. During the Gulf War, for the first time in history, the critical intelligence that is at the center of all successful military campaigns was gathered from outer space, making the last major war of the 20th century the very first space war.

The question of whether humankind will use outer space for military purposes has been rendered moot—we already do, and have been using space to wage war since the Soviets launched the first *Sputnik* in the 1950s. This trend is not likely to end. In fact, war in space is far more likely than at any time during human history, because satellite reconnaissance and communication have become critical to the proper function of a modern, high-tech military force.

During the Cold War, both the United States and the Soviet Union sought a means to locate and target their opponent's ballistic missiles so that, in the event of hostilities, the missiles would be destroyed before they were ever launched. For the Soviet Union, which had previously relied on humint—human intelligence gathered from spies—*Sputnik* was but the first stumbling step toward an effective system of reconnaissance satellites. Before *Sputnik*, the United States relied on high-altitude manned spy planes like the U-2 and its advanced Hycon-B camera to snoop on Soviet activities. Unfortunately for America, this practice backfired when a U-2 was shot down over Sverdlovsk, Central Russia, by an SA-2 missile.

Realizing that they had been trumped by the launch of *Sputnik*, while over-flights across the Soviet Union were now impossible due to Soviet advanced anti-aircraft technology, the U.S. military and the intelligence services began intensive satellite programs of their own. The result was *Corona*, a highly classified project began in 1959. *Corona* photo-reconnaissance satellites took pictures from orbit, and these photos were later dropped back into Earth's atmosphere using a capsule and parachute. As it drifted to Earth, the *Corona* capsule was captured in the atmosphere by an Air Force retrieval aircraft.

The effectiveness of first-generation systems like *Corona* was limited, because the two goals of reconnaissance are to cover the largest area possible while spotting the smallest objects. These twin goals could not be accomplished with the same optical and camera system. The *Discoverer* series of reconnaissance satellites launched in the 1960s solved this dilemma by installing two separate camera systems—the KH-4A wide angle system, and the KH-7 close-up ultra focusing system. But several problems remained. These satellites were limited to the amount of on-board film they were launched with, and there was a considerable time lag between the moment the photos were shot and when they got into the hands of analysts to evaluate them. Even with *Discoverer* and its sisters, America still had no real-time intelligence capability or early warning system in orbit.

The *Corona/Discoverer* satellites proved their worth during the Cuban Missile Crisis. After John F. Kennedy blamed the Eisenhower administration for causing a "missile gap" with the Soviet Union, Defense Secretary Robert McNamara poured over satellite data in an effort to learn where those elusive Soviet missiles were based. What he discovered was that *there was no missile gap*—that, in fact, the Russians had fewer than 100 missiles, and none that could reach the American heartland. In October 1961, during the Cuban Missile Crisis, President Kennedy could push Soviet premier Khrushchev to the "brink of nuclear war" because he knew Khrushchev was bluffing—the Soviets had no missile advantage and JFK knew it.

In 1970, the Defense Support Program launched the first of three real-time reconnaissance satellites to monitor Soviet missile launches. Placed in geo-stationary orbit 22,300 miles above the earth, these satellites were able to detect rocket plumes, and transmit data to several receiving stations scattered all over the globe and from there to NORAD (North American Aerospace Defense Command) headquarters in Colorado Springs, Colorado. Though effective, these satellites did not reach a level of sophistication of an ideal reconnaissance system. A perfect reconnaissance system should accomplish four tasks: target detection, target recognition, discernment of target characteristics (speed, altitude, deployment, munitions carried, and so on), and real-time results.

No satellite system in the world could perform all four of these tasks in the 1970s.

The next advance in technology came with the launching of the solar-powered KH-9 satellites, the first reconnaissance platforms designed to remain in orbit for an extended period of time, and the first to include a full-spectrum of intelligence data-gathering resources through the use of infrared film and two on-board cameras for enhanced stereo photography. The KH-9 was a quantum leap from previous systems, but the use of standard film and photographic techniques still presented problems. Photos produce an image that cannot be reduced to its constituent parts. In other words, sections of a photograph cannot be enhanced, sharpened, or expanded without suffering image degradation. Telescopic imaging was tried with moderate success, but a better system was necessary to receive sharp, clear, flexible images in real time.

The solution came from the world of cyberspace. Digitalization—breaking down an image into a string of data through the use of a computer, and then reassembling that image inside of earthbound computers—fulfilled the promise of an ideal reconnaissance system. Using digitalization, an image could be taken, beamed to Earth, reassembled by computers, and broken up into constituent parts that could be further manipulated and enhanced without significant degradation. The KH-11 satellite program begun in the 1970s and continuing throughout the 1980s launched the first space-based reconnaissance platforms to include digital-imaging systems. Though exact specifications are classified, we know that a KH-11 functioning during the Gulf

War had a resolution of about six inches (meaning it can detect objects as small as six inches in size), and later models doubtless have vastly improved capabilities.

Visual imaging is just one type of satellite reconnaissance. Imaging—even using the infrared spectrum—has limitations, so a means was sought to also use other parts of the electromagnetic spectrum to get a picture of what the enemy on Earth was doing. The result of this quest was LACROSSE, a radar imaging system. Radar is not affected by visible light, so a radar imaging system can be used night and day, through fog, clouds, haze, and the smoke of battle. Additionally, LACROSSE satellites *probably* contain infrared sensing, signal intelligence sensors, and other types of data gathering systems that are, of course, classified.

NO FOOLING AROUND WITH RADAR

Radar imaging has solved one of the great problems of aerial intelligence. During the two world wars, armies of all nations used decoys to fool enemy aerial reconnaissance and observation. During World War II, wood-and-canvas tanks and inflatable dummy trucks were positioned in England, opposite the coast of Calais, to convince the Nazis that the D-Day invasion would land there instead of Normandy.

Naked-eye observation and photographic intelligence can be fooled by decoys—radar imaging can not. Radar waves can detect the density of matter, and can easily distinguish between steel, cloth, rubber, or wood. A satellite with radar imaging and additional infrared sensors can also detect ambient temperature around an object and sense whether an engine has been used recently. With additional signals intelligence capacities that can monitor enemy radio transmissions, the same satellite could evaluate the amount and quality of enemy communications in a given region—decoys don't talk much; real soldiers do.

Satellites, using phased-array radar and synthetic aperture radar, as well as Doppler radar, have been launched to provide ever more accurate intelligence data to earthbound receivers. Beyond digital and radar imaging, the array of today's intelligence-gathering satellites include

ELINT (electronic intelligence) and SIGINT (signals intelligence) capabilities. These satellites use the radio-radar region of the electromagnetic spectrum to collect any communications signal stronger than background radiation. The first ELINT and SIGINT satellite launched by the United States in the 1970s was code-named RHYOLITE. The newest version of this type of satellite is code-named MAGNUM, the first of which was launched from the space shuttle in 1985.

Though the capabilities of MAGNUM are classified, we know that during the Gulf War the raw data provided by these satellites was so extensive that analysts had a difficult time keeping up with it. There were not enough human operators to review and evaluate all of the information pouring in, and to transform it into usable intelligence. Slow encryption and download times for visual imaging further slowed the intelligence gathering and evaluation process—enough so that General Schwarzkopf complained to Congress that he couldn't get the real-time intelligence he needed, despite these advanced, high-tech systems, because there were not enough human evaluators. There was also a pervasive and justified fear by the intelligence agencies (who operate most of these satellites) that by disseminating the raw data, the capabilities of our satellites might be revealed to the world.

Because of the many types of space-based systems put into orbit in the last 20 years, the United States possesses superb intelligence capabilities—perhaps the best in the world. Continued use of these platforms in the future will undoubtedly spur an enemy to try to degrade the satellite system or destroy it outright, which will result in warfare on a new battleground—the battleground of outer space.

Today, the United States depends heavily on space-based intelligence platforms for national defense. As we have seen, space capabilities are fundamental to the effectiveness of U.S. land, sea, and air forces. In recognition of this new reality, the United States Space Command (USSPACECOM) was formed in 1985. The Space Command, like the United States Special Operations Command (USSOC), is one of the nation's nine unified, multi-service commands. The components of

USSPACECOM are the Army Space Command in Colorado Springs, Colorado; the Naval Network and Space Operations Command in Dahlgren, Virginia; Space Air Force at Vandenberg Air Force Base in California; the Joint Task Force-Computer Network Operations in Arlington, Virginia; and the Joint Information Operations Center at Lackland Air Force Base in Texas. No U.S. government intelligence agency directly participates in the activities of the U.S. Space Command.

The men and women of Space Command put the satellites that provide military intelligence capabilities into orbit, operate them, protect them, and ensure that the information they provide is exactly what America's military requires. USSPACECOM coordinates the use of the Department of Defense's military space forces to perform specific tasks, including the following:

- **Missile Warning.** Defense Support Program satellites and ground-based radar systems provide strategic and theater ballistic missile warning to the government and to deployed troops worldwide.

- **Communications.** Communication satellites provide constant global connectivity with deployed forces.

- **Navigation.** The Space Command's Global Positioning System of 28 satellites provides precise navigation and timing support to coordinate the positioning and maneuvering of U.S. aircrews, naval forces, and ground forces.

- **Weather.** The Defense Meteorological Satellite Program collects and distributes global weather data.

- **Imagery and Signals Intelligence.** U.S. military space operators coordinate space-based imagery between intelligence agencies and planners within Unified Commands.

- **Space Support.** Launching and operating satellites, including telemetry and tracking. USSPACECOM launches take place at Cape Canaveral Air Force Station, Florida, or Vandenberg Air Force Base, California.

- **Force Enhancement.** Satellite communications, navigation, weather, missile warning, and intelligence.

- **Space Control.** Assuring American access to and freedom of operation in space, and denying hostile powers the same.

- **Force Application.** Researching and developing space-based capabilities that have the potential to engage adversaries from space.
- **Computer Network Defense/Computer Network Attack (CND/CNA).** Computer Network Defense includes protecting and defending information, computers, and networks from disruption, denial, degradation, or destruction. Computer Network Attack includes developing the capabilities to disrupt, deny, degrade, or destroy information resident in computers, computer networks, or even the computers and networks themselves. The USSPACECOM component command for these missions is Joint Task Force—Computer Network Operations.

USSPACECOM was headquartered at Peterson Air Force Base in Colorado until October 1, 2002, when it was moved to Offitt Air Force Base in Nebraska. The U.S. Space Command employs nearly 900 civilian and military personnel, and has an annual operations budget of about $67 million (fiscal year 2002). The establishment of the U.S. Space Command will someday be understood as a major event in the history of warfare. The creation of such a strategic force during a period of relative peace signals the tactical necessity of using outer space for reconnaissance purposes. It also signals the dawn of a new era in space-based military activities.

To wage a war in space, we must create a more efficient means to get material there. Rockets are fine, but require an enormous amount of money to transport a relatively small payload. The Space Shuttle was supposed to solve this problem, but it is estimated that it costs more than $9,000 a pound to move material into orbit in the shuttle—hardly as economical as the United Parcel Service! To move a conflict outside the atmosphere of Earth requires a better, more efficient, and economical propulsion system than a rocket engine.

The internal-combustion piston engine was the power plant behind the military aircraft of World War I and II. The jet engine powered military aircraft through the Cold War and into the 21st century. But it is the *scramjet* engine that will become the foundation for air power in the 21st century, once the problems inherent in designing and building an aircraft capable of ultra-high speeds are overcome.

A conventional *jet* engine draws air into its firing chamber, where it is compressed by turbines. Then combustible fuel is injected into that chamber and ignited. Rapid superheating of the compressed air causes it to expand out the back of the firing chamber, through the nozzles, to create thrust. A jet aircraft can go faster than a piston engine, but its speed is limited by the degree of compression—the greater the compression of air, the greater the thrust. But conventional jets have two major limitations: In a jet engine, the air is never compressed quickly enough to produce heating, hence the need for fuel and combustion; and jet engines cannot cope with the extreme heat caused by friction at speeds much greater than Mach 2.2.

A *ramjet* engine is a different matter. At Mach 2 or above, it encounters a wall of air that tries to enter its jet intakes at twice the speed of sound. At those speeds, a ramjet *moves fast enough to compress the air without the need for turbines*—in essence, the air compresses itself. This allows a ramjet aircraft to move up to six times the speed of sound. Above that speed, even a ramjet becomes inefficient—the shock of air rushing into the intakes begins to tax the engine and causes uneven compression.

To achieve speeds above six times the speed of sound, a *scramjet* engine is required. A scramjet allows air to enter its intakes at an oblique angle to minimize shock, uneven compression, and engine stress. The supersonic rush of air actually allows for a higher level of compression—hypercompression—so that when fuel is injected into this mass of hypercompressed air, it is done in two stages (which occur mere nanoseconds apart). First, fuel enters the mouth of the compression chamber to further increase compression, and then fuel is pumped into the chamber itself to induce expansion. This results in a phenomenal amount of thrust blasting out of the nozzles, permitting a scramjet aircraft to achieve speeds up to Mach 20 or 25.

The problem with scramjet engines is finding materials that can withstand the stresses of hypersonic speeds. Designers need to develop materials to form intakes that can handle such tremendous airflow, and a combustion chamber that won't rupture under such colossal pressure. Other stresses include atmospheric pressure and external heat created by air friction, as the aircraft races through the atmosphere 10 or 12 times faster than a bullet.

Presumably these problems can be overcome. In fact, the U.S. Air Force is betting on it. Several joint Air Force/National Aeronautics and Space Administration (NASA) projects are currently underway in a quest to develop a workable scramjet prototype. Because scramjet technology is relatively simple, fuel efficient (the ideal fuel is hydrogen), and even economical (presuming that the materials required for construction are somewhat reasonably priced), it is predicted that within a generation it will be possible to use an unmanned scramjet aircraft to deliver munitions onto an enemy at hypersonic speed and at intercontinental distances. Such a weapon would have a range in excess of 10,000 miles and a cruising speed of 8.5 miles per second.

Advances in computers would mean that this scramjet aircraft would have onboard sensors to receive instructions from other aircraft or even ground stations located a continent away. It could be preprogrammed to attack a specific target, or patrol in the general area of enemy forces until a target and its coordinates could be supplied by its controllers.

The dissemination of scramjet technology would have a huge impact on the geopolitical situation. Such a weapon could cross an ocean in eight minutes, or the continental United States in six. If the on-board computer and sensor systems were as brilliant as computers themselves soon will become, then such a weapon could shift the balance of power forever. And remember that scramjets and sensors and computers are all products of technology and innovation—even a small industrialized nation has the potential to develop such weapons. If Taiwan could build a smart scramjet platform, it could bring the People's Republic of China to its knees. If North Korea could build such a weapon, the United States could be threatened.

Mass armies are not required to deploy such technology—all that is required is a highly advanced industrial base. Nor can a massed army hide from such a weapon—if a scramjet is smart enough, or is precisely directed through the use of space-based, real-time satellite systems, then warships, bombers, tanks, and armies, could conceivably become a thing of the past. If you find the above hard to believe, then ponder this—according to many published reports, an effective scramjet may already exist. And according to some reports, at least one scramjet aircraft may have been tested in the mid-1990s.

During the Reagan administration, work was done on the National Aerospace Plane christened the X-30. The X-30 was to be a scramjet-powered passenger aircraft that would fly at speeds in excess of Mach 6. The proposed aircraft was to be about the size of a Boeing 727 with ultra-high altitude, edge-of-space capabilities. When Vice President George Bush became president, the promising NASP program was abruptly cancelled—but that's not the end of the story.

THE TOP-SECRET AURORA

During the early 1990s, several sources reported sighting a scramjet aircraft in flight—and at the oddest places. In December 1992, *Janes Defense Weekly* published an article by Bill Sweetman detailing a new hypersonic reconnaissance plane code-named Aurora, which he suspected was already a reality. Supposedly this aircraft (*Janes* provided sketches) was sighted several places—in the skies over Edwards Air Force Base, over the Republic of Georgia in the former Soviet Union, and even over the North Sea.

A separate article, written by William B. Scott for *Aviation Week and Space Technology Review* (August 24, 1992), reported that a United Airlines 747 crew claimed to have had a near-collision with the mysterious, top-secret aircraft. By combining these reports, we begin to get a picture of an aircraft that can travel at Mach 8 or above, and is said to "thrum"—emitting a beating or pounding sound instead of a continuous throaty roar—and to expel a series of oval smoke rings instead of straight-line contrail. If the Aurora exists—and there is as yet precious little evidence to suggest that it does—then the reality of such an aircraft would indicate that at least some of problems encountered in designing and building a hypersonic aircraft have been solved. If so, then the Aurora has paved the way for the hypersonic unmanned weapons platforms.

In March 1992, the Air Force issued a request for proposals for a high-tech project called the Hypersonic Aerodynamic Weapon Demonstration Program Definition (HAW). The Air Force wanted an aircraft that would "incorporate high hypersonic lift-to-drag rations, allowing long-range glides or extensive terminal maneuvers" with "guidance, navigation, control, sensors, warhead, fusing, and other applicable sub-systems appropriate for precision strike."[1]

This weapon would not be a true cruise missile, or an intercontinental ballistic missile (ICBM). It would differ from the Tomahawk in that it would not cruise low and slow, but would dive out of the sky high and very fast. It would probably also be a glide weapon—after initial acceleration from a scramjet booster, it would perhaps jettison the thruster and use its own aerodynamic lifting body to glide itself to the target. Although not quite the same advanced design just described, the HAW might be an interesting interim technology that will quickly lead to more promising—and more advanced—design innovations.

Out of the ashes of the "failed" Strategic Defense Initiative (SDI, or "Star Wars Defense") research, a phoenix may yet arise—several of them, in fact.

As an adjunct to SDI, the U.S. Air Force and the National Aeronautics and Space Administration (NASA) have ongoing research programs working to develop manned trans-atmospheric vehicles (TAVs) capable of taking off from conventional runways (or being launched from conventional jet aircraft). TAVs could reach Earth's orbit, travel through space, reenter the atmosphere, and land on a conventional runway. Such a vehicle would require a unique propulsion system—or several of them packed on top of one another. Under consideration to propel such aircraft are conventional rocket motors powered by more efficient and cheaper fuel, vacuum-thrust motors, cryogenic rocket motors, pulse-detonation wave engines, and the scramjet. Much of the technology for the TAV project was developed during the cancelled X-30 program.

Meanwhile, in 1996, Lockheed Martin's "Skunk Works" was selected to head the multi-corporation team of developers working on the project that superseded the X-30—the X-33 VentureStar Single-Stage-to-Orbit (SSTO) Reusable Launch Vehicle (RLV)—a space-shuttle-size vehicle capable of lifting a payload of up to 25 tons into outer space. The prototype is being built at the former B-1B bomber facility at Palmsdale, California. Problems with the rocket engines and several design flaws have placed this program in jeopardy, but a smaller unmanned design for an RVL—the Orbital Sciences X-34—made its first glide flight in 1999. With the addition of a proposed Fastrec liquid-oxygen/kerosene-fueled rocket motor, the X-34 will achieve speeds in excess of Mach 8.

More sophisticated than the X-34 is the proposed X-38 Ranger Crew Rescue Vehicle (CRV), which is scheduled to fly in 2005. This vehicle is intended to be a crew rescue ship for the international space station, but could also be used as a satellite support vehicle.

Even more advanced is the proposed X-43A, a manned hypersonic strike aircraft that would skip along the upper reaches of Earth's atmosphere, to reach any point on the planet in under two hours. The X-43A prototype is 215 feet long, wedge-shaped, and can carry nearly 100,000 pounds of munitions over a range of nearly 7,000 miles while cruising at Mach 10. According to proposed flight characteristics, the X-43A would take off from a 10,000-foot conventional runway, and then use scramjets to accelerate to Mach 10 while climbing to 115,000 to 120,000 feet, where the air-breathing engine would be shut down as the aircraft left the atmosphere. The aircraft would continue to climb to about 200,000 to 250,000 feet, using its own momentum, and then make a descent back to around 200,000 feet. Firing its engine every 20 or 30 seconds after that, the aircraft would quite literally surf through the air on its own shockwave, like a rock skips across a pond.

Another result of hypersonic research attached to SDI is the Super High Altitude Research Project (SHARP) gun, developed by a team at the Lawrence Livermore Labs. The SHARP gun is a low-cost solution to the problem of propellants and thrust engines. The problem with conventional rockets is that they are costly, mainly because rockets have to carry their own energy source—the engine and fuel—with them. A rocket is about 80 to 90 percent engine and fuel, and 10 to 20 percent payload.

The SHARP gun solves this problem through totally conventional means—a large space-bound projectile is fired through a tube exactly like a bullet is fired from a gun. In the SHARP gun, a mixture of air and methane is ignited, driving a piston down a tube and hyper-compressing hydrogen. The compressed hydrogen rushes into a second tube containing the projectile, which is forced out at a velocity of thousands of miles per hour.

Very much in the vein of Jules Verne's space cannon featured in *A Trip to the Moon* (filmed in 1958 as *From the Earth to the Moon*), the major problem with the SHARP gun is that living protoplasm cannot survive such a rapid and dramatic initial acceleration—a blast that results in more than 1,500 gs. This limitation didn't matter to the SDI program—the solid projectile in an SDI SHARP gun was to be made of high-density depleted uranium, and was supposed to destroy Soviet ICBMs in flight. But SHARP technology could also be used to launch equipment and cargo into orbit to re-supply the international space station, and do it much more cost effectively than the space shuttle.

The SHARP gun could also be coupled with scramjet and cruise missile technology to create an unmanned attack vehicle (UAV) that can be deployed anywhere on Earth in mere minutes. The SHARP gun could even be used to fire "smart artillery" to destroy an enemy thousands of miles away in minutes. After the initial pulse launches the "shell," it can climb into orbit, circle the globe, and drop back through the atmosphere at a predetermined target area, to annihilate it—without a single soldier, sailor, or airman leaving the continental United States.

More promising is the use of a cruise scramjet missile launched from a SHARP gun. In flight, such a weapon could discard its outer shell (or sabot) or use that shell as a reentry shield, and then deploy wings to make the platform aerodynamic once it has returned to the atmosphere. With the smart capabilities outlined previously, this platform can become an effective, instantly deployed weapons system for the 21st century. A SHARP-fired smart scramjet could be used to attack an enemy in the field, or strike at a target in the enemy's homeland without placing American soldiers or airmen in harm's way.

For the next few decades, any war in space will be a battle between satellites in orbit and ground-based rocket-propelled weapons fired into orbit to destroy them. We have seen that the satellite systems now in place are critical to America's success on the battlefield. In the next big war, or perhaps in one of the smaller conflicts now brewing, a belligerent power such as China, North Korea, or another nation-state with the tools and technology to launch satellites into orbit will—by necessity—have to destroy or disable America's satellite system to prevail.

Destroying a satellite in space is both simpler and more complex than it sounds. A rocket launched from Pyongyang or Beijing could rise into orbit, track down the noisy satellite (all satellites are noisy, emitting waves and radiation in many different frequencies and spectrums) and get close enough to it to discharge and disperse a cloud of shrapnel. As the satellite drifts through this cloud on its predetermined orbit, it will be destroyed.

Indeed, three types of satellites would make rich targets for any nation hostile to America: optical and radar reconnaissance satellites in polar and elliptical low-earth orbit, Navstar satellites in circular orbit, and ELINT and SIGINT satellites in both geo-stationary and medium Earth orbit.

Eliminating or crippling any of these satellite systems would severely damage America's ability to wage war. Beyond attacking these satellites directly with ground-based rockets, there are other ways to cripple or destroy America's space-based reconnaissance and intelligence capabilities. An enemy could attack American ground-receiving stations, for instance, *if* they could locate them all and mount an attack on such installations, both outside of and within the continental United States. An enemy could also attack the satellite systems in space using high-energy beams, but more likely such an attack will be made with electronic-warfare techniques, used to disrupt, corrupt, jam, or supplant the satellite-to-receiver data flow.

To avoid or elude attack, the next generation of intelligence-gathering satellites will most likely possess the ability to maneuver in radical and unpredictable patterns. There may also be a system of passive defenses applied to future satellites—perhaps some form of space-based stealth technology, or the space-based equivalent of chaff dispensers, to fire some anti-sensor or sensor-fooling material to draw fire away from the real target. Or perhaps future satellites will simply wear composite armor plating to protect the delicate internal mechanism from damage.

Eventually, a system of active defenses designed to destroy anti-satellite weapons will most likely be developed and deployed. The intelligence and reconnaissance satellites themselves may carry defenses, or they may be accompanied by "escort satellites" that will do the fighting for them. Probably a combination of these strategies will be used, the most important of which is maneuverability.

Already satellites contain propulsion systems. Some embody a new technology called electric propulsion. Electrical energy is converted from solar energy and is used to excite a propellant—ammonia and hydrogen have been used, but recently the Japanese have developed a technique to ionize xenon, which can also be used as a propellant. Nuclear power is being used right now in Russian satellites, and it is believed that the American KH-11 reconnaissance satellites have a Bus-1 nuclear propulsion system as well. All of these propulsion systems have significant drawbacks. Electrical propulsion engines work slowly—too slowly to be effective in case of attack. Nuclear reactors are large and heavy, and take up much room in a small reconnaissance satellite, so nuclear thrust is also imperfect. Ultimately, some form of satellite defense must be found, because anti-satellite weapons will soon appear to challenge America supremacy in this highest of frontiers.

Currently, international treaties and agreements prohibit the positioning of space-based weapons. In time, it is likely that such treaties will be ignored or flaunted—perhaps by the United States. The reason for this is that all of the technology just described is limited because all of it has to battle against the force of gravity—hypersonic, SHARP guns, ground-to-space missiles, and so on, are all limited by the constraints of celestial physics.

But if weapons were actually based in space, in manned or unmanned platforms placed in high circular or elliptic orbit, then gravity actually would become an ally. Kinetic weapons need only be aimed and fired—after that, their own inertia and the Earth's gravity will pull them down to their targets.

Laser beams, particle beam, microwave, or solar-powered guns fired from outer space to targets below are not beyond the realm of possibility, once the thorny issues of power and atmospheric interference are solved.

In time, the wars in space will resemble the wars formerly waged on Earth's oceans—with small manned space ships and large manned space platforms battling one another to gain supremacy in the highest frontier of all—and by extension all of the Earth below.

One thing is certain—despite treaties and agreements, good intentions and international cooperation, humankind will inevitably carry their hostilities and hatreds, their grievances and greed, their belligerence and bias, and their dreams of lasting peace through superior firepower into outer space. One day—whether sooner or later—the first human war beyond the bounds of Earth's gravity will be fought.

PART 5

IN A LEAGUE OF ITS OWN: NONLETHAL WEAPONRY

"First do no harm."
—The Hippocratic Oath

"Cruelty in war buyeth conquest at the dearest price."
—Sir Philip Sidney

CHAPTER 11

TAKING KILLING OUT OF WAR

As the world's only superpower in the 21st century, the United States has also become the world's police force. This is inevitable. In the past half-century or more, beleaguered nations have relied on the American people as a beacon of hope in the darkest of times and, generally speaking, America has responded to their need, repeatedly deploying troops and risking American lives while asking for precious little in return.

Despite charges of "American Imperialism" by critics regarding America's role in World War II, the Berlin Airlift, the Korean War, the Vietnam War, and in the Persian Gulf, the people of the United States have repeatedly put their own lives or the lives of their sons and daughters in harm's way to protect the ideals of freedom, self-determination, and prosperity around the globe.

In the 20th century, the world learned to rely on American authority, and has grown dependent on American military might. But peacekeeping in the chaos of the last 50 years has become a full-time job, and a risky one at that. Sometimes the peoples assisted, those who were provided with care and nutrition and humanitarian aid of all sorts, are convinced by despots and warlords to turn on their benefactors. This happened in Somalia— with tragic and bloody results.

The purpose of peacekeeping and humanitarian aid missions is to help, not harm the threatened populace, but at times that is impossible. Faced with the threat of violence, a nation can do one of two things— withdraw from peacekeeping and humanitarian aid duties, or fight back— an action that will ultimately result in more death and suffering, this time on both sides.

But now there may be a *third* option. Currently, a vast array of non-lethal weapons and systems are being developed for military uses. Some of this technology has been around for a long time, but is finding new military applications in our savage new times. Some of this technology has been used previously by law-enforcement organizations in the United States and in other countries. And some of this technology is so innovative and cutting edge that we have not seen its like before.

Imagine how many appalling and tragic incidents could have been avoided through the use of nonlethal technology. How would the Battle of Mogadishu have been different if the Army Rangers and Delta Force could have sprayed the entire Bakara Market with a nonlethal area neutralizer? Instead of a night-long running battle with Aidid's technicals and an enflamed local population of misguided Somalis—a 14-hour fight that resulted in hundreds of casualties—the men of Delta Force could have moved in *after* all the Somali citizens inside of a 5-mile radius had been rendered unconscious. They could have located and snatched up the warlord and his followers. And a few hours later, the people of Mogadishu would awaken with a mild hangover, to find themselves liberated from the iron-handed rule of the tyrants who starved and oppressed them.

How would the national shame that was the Siege of Waco have ended differently if the FBI and ATF had used an area neutralizer on the Branch Davidians? There would have been no fire, no massacre, and no one to question the motives and violent tactics of the Justice Department, the administration, and federal law enforcement agencies.

Though these same law enforcement agencies, along with the military, have been slow to grasp the usefulness of nonlethal technology in myriad dangerous situations, they are slowly catching on. Even now, with limited interest among the military's sometimes hidebound top brass, nonlethal technology is being called the fastest-growing type of military and law enforcement technology now in existence. Only time will tell if such technology will become deployed regularly, along with the other weapons in America's arsenal.

Seven known types of nonlethal technology are currently available to the U.S. military, and there may be more that are still classified. Some of these technologies are not new, and with some, the strategies behind them are very old indeed. All of these types of nonlethal technology are viable options to killing the enemy and drinking beer from their skulls. The seven types of nonlethal technology currently available are biological, informational, acoustical, low kinetic, restraining, chemical, and electromagnetic. Each of these systems have workable examples in use or available for deployment right now and other systems still in the developmental stage.

One of the oldest forms of warfare is *biological*. In ancient times, Roman armies contaminated wells with dead animals to deprive the enemy of potable water. In the middle ages, cities under siege were often bombarded with the disease-infected corpses of humans and animals in an attempt to cause epidemics among the besieged populace. Today, severe international treaties restrict the use of biological weapons—and rightly so. But the nonlethal biological weapons we are speaking of here do not attack the human organism. Rather, they attack the materials used in modern warfare.

Research is ongoing to create a biological organism that ingests concrete or asphalt. Bio-engineered with a suicide gene to limit its spread as well as its life span, these microbes could degrade runways, roadways, and structures—diminishing a technological nation's ability to wage war effectively. Microbes are also being bio-engineered to ingest plastic and rubber to destroy tires and seals—and even composite materials. A rubber, plastic, or composite material-ingesting microbe could degrade the radar-absorbent characteristics of a stealth aircraft by eating away at its coating, or short out electrical systems by degrading the insulation around wires and circuits. Organic materials are also being considered that degrade explosive material, so that when it is detonated it produces less than a fourth of its potential power; and bio-organisms that eat copper wire and silicone are also being considered, or are actively being developed.

BIOHAZARDS AND TOXIC SHOCKS

Biological warfare is the military use of disease-causing bacteria, viruses, rickettsia, or fungi. Closely aligned with biological warfare is *toxic* warfare, the use of poisonous substances naturally produced by living organisms. Both types of weapons are currently prohibited by international law.

During the Kennedy and Johnson administrations, the United States produced and stockpiled 13 biological and toxic agents—including smallpox, anthrax, botulin, and tricothecenes (fungal toxin agents)—reaching peak production in 1965. In the late 1960s, the U.S. military determined that biological weapons were unreliable, and they were unilaterally banned—along with chemical agents—by President Richard M. Nixon in 1969.

In 1975, the rules of the international Convention on Biological and Toxins entered force. Unfortunately, adherence to the precepts of this convention is based on mutual trust—there are no verification or inspection procedures built into the agreement. Though several nations, including the United States, have ongoing research programs involving biological agents, because such weapons are unreliable and very difficult to control, it is doubtful that any civilized nation will ever deploy them. The use of emerging nonlethal biological weapons, such as rubber-eating bacteria, also violates the Biological and Toxin Convention.

Some nonlethal technologies *do* have an effect on the human organism. Working in conjunction with the Pentagon, the Monell Chemical Senses Center in Philadelphia, Pennsylvania has formulated a variety of smells that are so repellent that they can quickly clear a public space of rioters, enemy forces, or anyone who can breathe. Researchers have tested the effectiveness of such odors as vomit, rotten eggs, burnt hair, sewage, decaying flesh, and a vile chemical mixture known as *U.S. Government Standard Bathroom Malodor*.

The problem with biological agents like malodorants is the same problem with all biological weapons—once they are unleashed, they are difficult to control. Such weapons are also controversial, and some of them may even be illegal. Like the situation with chemical nonlethal weapons such as faster-acting, weaponized forms of antidepressants, opiates, and even so-called "club drugs" like ecstasy, which could be administered to unruly crowds through an aerosol spray, the use of malodorants will most likely be deemed illegal under international law—even though the use of substances like tear gas and pepper spray are perfectly legal. If the only alternative to the use of these chemical irritants is slaughtering members of an unruly mob (and violence sometimes *is* the only alternative), then we should at least consider the nonlethal options that may be available—even if such innovations make human rights organizations like Amnesty International uncomfortable.

Informational warfare (IW)—or propaganda, both black and white, along with misinformation—have been with us as long as warfare … and political parties. It's a way of waging war that relies not on bombs, bullets, or guns, but the dissemination of information and ideas. In the context of war, propaganda is information—printed, broadcast, or announced—that originates from the enemy camp and is recognizable as such. Information that seems to originate from a trusted source, but has in reality been planted by the enemy, is called black propaganda because it appears to be authentic even though it is not. "White" propaganda is propaganda straight from the source. Radio show hosts Tokyo Rose and Axis Sally, two examples of white propaganda, played music interspersed with statements about the futility of Allied efforts in the face of Hitler's and Tojo's undefeatable hordes, which was then beamed to Allied troops. The thousands of fliers dropped on populations during wartime, urging them to surrender because resistance is futile, is also a form of blatant, or white, propaganda.

Computer hacking is another form of informational warfare. Viruses, computer worms, Trojan Horses, stealth codes, and bombs can negatively affect computer systems and can be launched through hardware, software, or firmware. Infection can be induced through any vulnerable point in the system—through telephone lines, disks, other programs, the power grid, servers, or even a local area network that is hacked or sabotaged. This aspect of informational warfare is exploding exponentially, and no nationwide computer networks are adequately protected from computer attack.

MISINFORMED FICTION

Misinformation is another form of informational warfare. A good example of this appears in Tom Clancy's novel *Debt of Honor*. During a time of heightened tensions between the United States and Japan, Boeing *appears* to issue a statement that the autopilot on several types of aircraft may malfunction on landing, and advises pilots to disable this function during approach. Later in the novel, as a pair of expensive Boeing Japanese AWACs are landing in mainland Japan, two CIA agents beam the cockpits with blinding laser light from a hotel room close to the military airfield. Both planes crash because the autopilot function that could have prevented the airplanes from crashing were disabled—as per the manufacturer's instructions. But the directive from Boeing was actually a piece of misinformation distributed by the CIA, so they could have a means to destroy the Japanese AWACs covertly, without an act of war.

Perhaps the best known application of *acoustics* in warfare was Joshua and the walls of Jericho. According to the Old Testament of the Bible, a single blast from seven trumpets caused the walls of the besieged city to crumble in the first recorded example of *sound as a weapon*.

In the 1990s, a primitive use of acoustics in warfare was made to drive Manuel Noriega out of the Papal Nuncio in Panama City during Operation Just Cause. Huge loudspeakers were placed around the compound where the fugitive dictator had taken refuge from American forces. In respect for the ancient concept of sanctuary, U.S. forces did not storm the church. Rather, around-the-clock music was directed toward the Nuncio. The loudspeakers blasted the most obnoxious heavy metal music imaginable, until neighbors, and even the Vatican, complained. Eventually, Noriega surrendered, and though this method did not have a decisive effect on the outcome of the operation, this tactic might have been much more effective had more grating sounds been used.

Contemporary nonlethal weapons research is concentrating on three levels of acoustic frequencies—infra-sound, audible sound, and ultra-sound. Infra-sound is at the low end of the wave spectrum, and not generally audible to human ears. Though not the most dangerous form of acoustical irritant, continued exposure to low energy infra-sound for extended periods of time can cause illness in humans and damage to buildings and structures, and such sounds are quite difficult to screen out. Higher intensities can induce nausea and disorientation, and may be more effective for use as a nonlethal weapon.

Audible sound is less promising, but experiments to induce physical and psychological effects are now being conducted. Audible sound in nonlethal weapons will work much like long-term exposure to loud rock music. Rock concerts often exceed 110 decibels (dB). Continued exposure to any sound at that level can result in a pronounced hearing loss. For a few seconds, some rock concerts peak at around 130 dB, which can be quite harmful. Generally, researchers agree that any frequency above 150 dB can cause damage to the human body's internal organs as well as to eardrums. The problem with sound at this level is that it affects everyone in the immediate general vicinity—friend or foe.

It is in the realm of ultrasound—in frequencies above 20 kilohertz—that the most promising research is now being conducted. Already in use as a diagnostic tool, ultrasound has also been found to induce a noticeable heating effect on organisms with prolonged exposure. Right now, experiments that will lead to the creation of an anti-material acoustical weapons—a noise that will knock down a wall like the ones surrounding Jericho—are ongoing. Research on such a weapon has been conducted in the former Soviet Union and continues in the United States—all of it highly classified. In the not-too-distant future, a building or aircraft may be shattered by an ultrasound projection weapon in the same way that a gifted opera singer can shatter a wine glass.

Low-kinetic energy, the most popular variant of which is rubber bullets, has been around in some form for about 35 years. Developed by the British for use against the rebellious population of Northern Ireland, the use of rubber bullets has spread throughout the world. The first rubber bullets were fired from a gas-grenade launcher and were relatively safe when striking the lower body or limbs—but could do harm or result in fatalities if the strike was made to the face, throat, or head. Early rubber bullets left the barrel at about 200 miles per hour and were usually fired within a range of 25 to 30 yards.

In response to the Intifada in the West Bank and Gaza Strip, the Israelis developed a number of low-kinetic weapons for use by the Israeli Defense Force. The best of these systems is the MA/RA 83 and MA/RA 88—both designed to attach to a standard M16 rifle. The non-lethal munitions containers can be fitted on the rifle in five or six seconds and are fired using a 5.56mm ballistic cartridge. The system can also be removed from the rifle quickly if the use of lethal force—real bullets—becomes necessary.

In the United States, researchers have developed nearly 70 forms of nonlethal, low-kinetic energy munitions, including various types of bullets and grenades. One promising development comes from the Accuracy Systems Ordinance Corporation of Arizona, which has invented a class of munitions that is designed to limit collateral damage in a terrorist or hostage incident. These munitions, called Special Purpose Low Lethality Anti-Terrorist (SPLLAT) shells, feature a ceramic slug that disintegrates upon impact—eliminating the danger of ricochet in a confined space.

Despite these precautions, the most common types of low-kinetic weapons now in use worldwide continue to pose a danger. That is why the Nonlethal Technology Innovation Center (NTIC), a research group located on the campus of the University of New Hampshire in Durham, is on a quest to design a softer, flatter bullet, including beanbag and sponge projectiles that spread out the impact and hit more like an open-handed slap than a punch to the gut. The NTIC is also working on even more radical technology—bullets that can be adjusted in the field to be harder or softer as the situation warrants. It is hoped that such munitions will provide flexibility to military and law enforcement personnel within the next decade.

Restraining technology is another form of nonlethal weapons technology. The first form of restraining technology widely available were the spikes sown in the path of vehicles involved in a high-speed chase. The fugitive driver races over the spikes, which promptly shreds the tires to ribbons, and the car is soon forced to stop.

Today, a wide array of nets and restraints are also being developed. A Massachusetts company called Foster-Miller has created the WebShot, a 10-foot wide Kevlar net packed into a cartridge and fired from a specially designed shotgun. The WebShot can entangle targets as far away as 25 feet, and larger nets are being developed to stop or restrain even bigger targets. The Portable Vehicle Arresting Barrier (PVAB), developed for the Pentagon by General Dynamics, is a nearly impervious, elastic web that springs up from the ground in an instant, to block a road or gate. This net can stop a 4-ton pickup truck traveling at 40 miles per hour, engulfing the vehicle and trapping the occupants inside.

A different kind of nonrestraining technology that nevertheless prohibits movement has emerged out of the chemical research done on nontraction and anti-traction agents. The Southwest Research Institute in Texas has created a gel for the Marine Corps' peacekeeping activities that is especially effective. This substance is so slippery that it is impossible to drive or even walk on any surface covered with it. Packed in a aerosol spray can or barrel, one shot on a door handle makes it too slippery to turn, and a coating on a road makes it quite literally impassible. This anti-traction gel is composed mostly of water, so it is nontoxic and biodegradable and dries completely in about 12 hours.

Most controversial is the research being done to develop *chemical* nonlethal technology. The chemical option is least attractive because it is a political hot-button issue. Chemical weapons are classed as weapons of mass destruction (WMDs), and even though something like tear gas or pepper spray is a long way from deadly Sarin or mustard gas, few governments or human rights organizations are willing to make such distinctions.

Two types of nonlethal chemical agents are being developed for potential military use—anti-material chemical agents, and anti-personnel chemical agents. Anti-material chemical agents include substances that will damage or destroy vehicles, engines, weapons, and munitions. So-called super-acids fall under this category.

Based on a combination of fluorine compounds, super-acids are up to a million times more damaging than hydrofluoric acid, which has the ability to eat through glass. Using a small pronged device called a caltrop that acts like a hypodermic needle, super-acid can be used to disable a vehicle. Similar to the spikes thrown on roadways in advance of a high-speed pursuit, these caltrops have an added punch—they are filled with a super-acid compound that not only flattens the tires but also chemically melts them, and melts the steel wheel the tire surrounds, too. Catastrophic damage results from using acid-filled caltrops as a delivery system, not only to vehicles, but to personnel who attempt to remedy the problem—rendering super-acids less than perfect in their "nonlethal" role.

A safer approach to attacking tires is with de-polymerization agents. Research at the University of Florida has previously focused on increasing the strength of polymer bonding to extend the life of a tire. But work done with catalytic agents that decrease or destroy polymer bonding have yielded dramatic results. With such an agent, delivery could be done with much smaller devices because a much smaller amount is necessary. Even better, because human beings are not made of polymers, the catalytic agent is nonlethal and harms only what it was invented to harm.

Another approach to disabling vehicles and aircraft is through the use of tiny, extremely abrasive ceramic or Carborundum particles that can be sucked into an engine through its air intake. These substances are so tiny that they are quickly distributed throughout the mechanism, to grind away at moving parts. The results are not immediate, but they are ultimately effective. The most promising way to cause an engine to ingest these substances is through an explosive burst of metal or ceramic shards. Igniting metals like cerium oxide produce extremely fine ceramic dust in particles so small that they can even penetrate most engine filters.

Fuels are also vulnerable to nonlethal chemical attack. Direct viscosification agents may be added to fuel to make it impossible for it to be aerosolized inside the engine. Such contamination is difficult to detect, but once the fuel is inside the tank it becomes gelled, to clog fuel pumps and choke carburetors. Super-lubricants like Teflon spray or potassium soap (mentioned previously) are considered nonlethal chemical weapons, and can be used to effectively halt vehicle and troop movement through a contaminated area by coating the roadbed and making it too slippery to touch. Researchers are also busy developing *electromagnetic* nonlethal weapons. Using a tight, focused beam of energy that flash-heats its target from a distance, so-called directed energy beams do not burn or harm flesh. Rather, they create an unbearably painful burning sensation in human skin tissue by stimulating the nerves. In the last five years, the Air Force Research Laboratory has spent $40 million developing a Humvee-mounted directed-energy weapon that will enter front-line deployment by 2009. This nonlethal technology will work to quell riots and disperse crowds without lethal consequences.

More ominous is the development of eye-dazzling and eye-damaging directed light or laser weapons for battlefield use. These weapons can cause permanent and temporary blindness and, though nonlethal, they are surely dangerous. Unfortunately, countermeasures against such optical weapons are not easy to develop because of the various optical sensors and sight-enhancement systems currently in use by military forces worldwide. Directed light weapons can overwhelm the light filters in many types of night vision gear, which could potentially harm users. Though these devices may shield the vision against some frequencies, none of them are effective against all frequencies, and the use of such combat lasers has the potential to cause permanent damage to its victims.

The effectiveness of electromagnetic pulse (EMP) weapons is widely known. The basic concept of an EMP weapon is to generate one or a series of very intense pulses of electromagnetic power to penetrate electrical devices, such as computers and power generators, and destroy sensitive circuitry. Nuclear bombs detonated in the air cause an electro-magnetic pulse that can fry communications systems, power grids, and computers for miles around the blast area. An electromagnetic attack leaves equipment burned out as the electrical surge races through the power cables or overloads a computer terminal. However, a nuclear detonation is not required to create an electromagnetic pulse—there are far more effective and less catastrophic means available to the military—most of them classified. EMP bombs, jammers, and projectors are currently being developed to simulate the electromagnetic effect without the atomic detonation.

The effectiveness of EMP weapons has become a great concern to the military. Shielding equipment has only limited effectiveness against EMP weapons, yet increases the cost and weight of such systems exponentially. Because commercial and personal computers and equipment remain vulnerable to EMP attack, all advanced industrial societies have the potential to sustain major damage, should an attack occur.

The Mission Research Corporation of Santa Barbara, California, is working on a pulsed-energy projectile (PEP) that superheats the surface moisture around a target so rapidly that it literally explodes, producing a bright flash of light and a loud bang. The effect is similar to a stun grenade, but, unlike a grenade, the PEP travels at nearly the speed of light and can take out a target with pinpoint accuracy.

Stun guns—or tasers—have been used by law-enforcement organizations and private citizens for more than two decades. One new system, called the Air Taser, uses compressed air to fire an electrical dart that is aimed by a laser-targeting system. Powered by a 9-volt battery, the air taser delivers a 25,000-volt shock that causes a temporary loss of neuromuscular control. Another much more effective type of taser is currently under development—a flashlight-shaped weapon invented by HSV Technologies. Like conventional tasers, this device transmits a powerful electric current. Unlike conventional tasers, the electricity does not flow through a wire, but along a beam of ultraviolet light. To use the system, one needs only shine that light on a human target—the result is a wireless taser that can temporarily paralyze targets as far away as a mile.

The development of these nonlethal technologies has attracted negative attention in recent years, drawing fire from human-rights organizations and political activists worldwide, who contend that such technology is actually being created to quell dissent. These same individuals and organizations maintain that most of the nonlethal weapons currently being designed actually violate international law.

Though some of these arguments have merit—for we cannot say with any certainty that countries won't use such technology to suppress their own dissenters—the potential of nonlethal technology to create a kinder, gentler military presence exists now, and should be exploited.

Human-rights activists may complain, but one must then pose a simple question: How can a war waged with nonlethal weapons—a war in which the enemy is restrained, confused, disoriented, rendered unconscious, or captured, but not killed—be branded *inhumane?*

CHAPTER 12

CONCLUSION: WHEN, WHERE, AND HOW TO DEPLOY?

The world is a more dangerous place than it has been at any time in human history. This is not hyperbole, but fact. Despite Western affluence—and the comfort and relative security it buys—most of the world's population is sinking into depths of poverty, violence, and despair from which they can never hope to emerge. The result is a widespread, hopeless desperation that is a breeding ground for envy, aggression, political repression, religious fanaticism, genocide, and indiscriminate violence. Indeed, the levels of violence across the world are likely to rise in the coming decades.

Of course, humans have always lived in a violent and dangerous world. In the past, human aggression found an outlet in an endless stream of warlike despots, from Attila the Hun to Genghis Khan to Napoleon Bonaparte to Adolf Hitler. But Attila and Napoleon never possessed the technologies of mass destruction available to nearly all technologically advanced nation-states in the 21st century. While madmen throughout history—Caligula, Vlad Tepes, Adolf Hitler—may have had the *will* to destroy the world, they lacked the *means*. Today, weapons of total annihilation are proliferating, along with the other technological fruits of modernity. This ominous trend is also likely to continue well into this new century.

War in the 21st century will be conducted very differently than the conflicts of the past were. In previous centuries, wars were fought between nation-states on a relatively equal economic and technological footing—France vs. Britain, Britain vs. Germany, the United States vs. the Soviet Union, and so on. In this century, that is unlikely to be the case. Instead, wars will be waged by advanced industrialized nation-states against vast impoverished populations from failed pre-modern, emerging states—or even against primitive cultures that have been indoctrinated by fundamentalist religious fanaticism, or steeped in resentment toward more affluent and industrialized nations. One thing is certain: The nation-state against nation-state conflict will be the exception rather than the rule in the 21st century.

Yet the U.S. government and military sometimes seem reluctant to make basic strategic changes necessary to fight this new kind of war. See how quickly the administration of George W. Bush focused on Iraq after the events of September 11, 2001—even though there was no solid evidence at the time to suggest that Saddam Hussein had anything to do with the terrorist attack on the World Trade Center. Indeed, within days, the American people knew that the attack had been perpetrated by Osama Bin Laden and the al Qaeda terrorist network. Even then, the Bush doctrine came to revolve around an "axis of evil" that included four nation-states—Iran, Iraq, Libya, and North Korea—under the theory that such powers "sponsor" terrorism.

Could it be that we as a nation are more comfortable making war on another organized nation-state that mirrors our own? Is it possible that we are as yet psychologically and militarily unprepared to battle an elusive, stateless international entity like al Qaeda?

If this is the case, then we will have to change, and change quickly, for in this century there will likely be fewer conventional wars. The majority of our future conflicts will probably be asymmetrical and non-traditional. International terrorism and international crime syndicates will hurt our citizens and harm our country far more than hostile nation-states are likely to, and far-flung civil wars will intermittently threaten our national interest right here at home. Insurgencies—including criminal insurgencies like those perpetrated by drug cartels and the Russian mob—will continue to spark violence, terrorism, and revolt in places like Colombia, Chechnya, and the Golden Triangle of Asia. Cultural and religious warfare is likely to increase despite international humanitarian intervention, because in such conflicts it is virtually impossible to appease the belligerents without the expulsion or eradication of one side of the equation. Human-made and natural disasters will also demand a humanitarian response that will too often become a military occupation—witness the misguided UN attempts to alleviate the famine and violence in Somalia.

Such nontraditional conflicts will demand our attention in the present century. As a result of these pressures, the United States will most certainly wage wars in distant lands to protect the nation's vital economic interests, and the interests of the West and East Asia. It will also become embroiled in localized civil and fractional wars, and missions of humanitarian assistance in places like Africa, the Middle East, and South America. But given the finite resources of our military and the American tax dollars, politicians and strategists will be forced to weigh the impact on our nation before intervention is feasible. That means that we must collectively decide if a situation, no matter how tragic the human cost, is truly worth U.S. military deployment and the likely death of American citizens.

Although technology will provide us with amazing new weapons—many of which have been described in this book—with which to make war, it is more important that America's political leaders understand when it is appropriate to fight, and when it is more prudent to stand aside and do nothing, beyond watching the violence and chaos unfold on CNN.

Who polices the world? When is it wise to intervene, and when is it wiser to let the horrors the emerging world is certainly doomed to experience play themselves out without intervention?

These are questions that we as the citizens of the remaining superpower must ask ourselves. It is a problem that is likely to challenge our moral fiber and our national will and our national identity in the coming century. There are always good reasons to fight wars—and equally good reasons to stay out of them. Traditional reasons to make war are still valid. If a nation is attacked, for instance, it is wise and prudent to respond with force. But few mission profiles emerging in the 21st century will be so clear, or so clear-cut.

So when do we make war? Here are the sensible guidelines the Pentagon seeks to follow in the 21st century.

If we are attacked. War is most certainly justified in the case of a military or terrorist attack against U.S. citizens, military, or civilian installations, or property. No nation-state, no matter how powerful, can retain its national sovereignty, its national identity, or its respect within the international community if it doesn't respond in kind to a direct military or terrorist attack.

But what *is* an attack? Future provocations will most likely be asymmetrical in nature—there will be no more Pearl Harbors—that is, overt military attacks by hostile nations that result in major conflicts or even world wars. As Americans, we recognize that Pearl Harbor was an attack, that Iraq's invasion of Kuwait was an attack—and so was September 11.

But so, too, was the terrorist suicide bombing of the USS *Cole* in Yemen in 1999, as was the bombing of the U.S. embassies in Kenya and Tanzania that killed 236 people, including 12 Americans. The assault on U.S. Army Rangers engaged in a humanitarian aid mission in Somalia in 1993 was also an attack on American interests, and so was the first World Trade Center bombing that same year. In fact, each of these attacks was actually an escalation of the previous event, and all were connected. The fatal blunder of the Clinton administration was in refusing to recognize these attacks *as* attacks, and to treat them as "criminal events" rather than what they were: outright acts of war.

The first World Trade Center bombing in 1993 was a calculated terrorist attack carried out by foreign nationals within the continental United States. When the United States failed to respond in a timely or aggressive manner to that unprecedented attack, Muslim extremists pushed the envelope with each subsequent event, to see just how much they could get away with.

We know now that Osama bin Laden was instrumental in providing rocket-propelled grenades, training, and assistance to Farrah Adid's "technicals" in Somalia in October 1993. The result was the downing of two U.S. Army Black Hawk helicopters in Mogadishu, and the subsequent Battle of the Black Sea immortalized in Mark Bowden's book and Ridley Scott's film *Black Hawk Down*. Once again, the Clinton administration did nothing, beyond pulling out American troops and conceding defeat—a disastrous political move that further emboldened America's enemies.

Next came the bombings of the American Embassies in Kenya and Tanzania on August 7, 1998. Instead of deploying the military, the Clinton administration sent in the FBI, treating the incident as a criminal matter. Two years later, the USS *Cole* taking fuel in Yemen—a nation that is considered an ally of the United States—was bombed, with the loss of more than 70 American servicemen. Islamic extremists claimed responsibility for the bombing, but still the sitting president did nothing more than send in FBI agents—a tepid response that further emboldened our enemies.

Finally, al Qaeda terrorists hijacked two commercial airliners and smashed them into the World Trade Center and the Pentagon, killing more than 3,000 Americans. The government, under the George W. Bush administration, at last framed an appropriate response. However, if the previous administration had recognized the first attack as an act of war—not mere criminality—then the three subsequent events, and the loss of American, Somali, and Afghani lives, might have been prevented. To avoid such horrible blunders in the future, not only must the United States provide a swift and measured response to every terrorist act, but the government must also broaden the definition of *attack* to include *any act of aggression against the nation, its military, its business interests, or its citizens*. This should include attacks on U.S. citizens overseas, and attacks on our domestic or international infrastructure—including our national and international computer and telecommunications systems and their facilities.

This certainly doesn't mean that the United States should go to war every time a lunatic or malcontent sets off a bomb in a Paris McDonalds, a Berlin nightclub, or an Egyptian tour bus. Nor should we overreact every time a precocious Hong Kong hacker wreaks havoc on one of our computer systems. But the U.S. government must take any attack seriously enough to launch a measured military response if necessary, or see to it that the U.S. justice system or a foreign court metes out justice against the perpetrators.

Which brings us to another valid reason to deploy American troops— *to prevent future terrorism*. It is not enough to react to terrorism in the 21st century, the U.S. military and intelligence services must also be capable of acting first to prevent future acts of terrorism. If an American office is attacked overseas, the consequences can be tragic, but not devastating. But terrorism against U.S. citizens using weapons of mass destruction—biological agents, gas, dirty, or traditional nuclear weapons—have the potential to kill thousands, or even millions.

As a nation, we must be prepared militarily and psychologically to take decisive action and make a first strike against any group or nation that deploys such weapons against the United States or its allies. Though the concept of a first strike seems inimical to our national character, refraining from such action in the future will be impossible unless we are willing to suffer massive casualties in order to retain our "high moral ground." Deployment to curtail terrorism may mean a first strike against a remote terrorist compound in the deserts of the Middle East, or sending in U.S. special operations forces to wage a covert war, or taking down and perhaps occupying an entire nation-state that is a proven sponsor of terrorism.

First strike may also mean cutting off the head of a terrorist network or nation through all-out attack or even assassination. Before we can do this, the United States must collectively question the absurd and outdated ethos that permits us to maim and kill thousands of faceless, nameless, and probably innocent civilians without question, but forbids the destruction of a single individual despot to end a reign of terror.

There will, of course, be political resistance to such seemingly reckless first-strike military actions or assassinations from elements both inside and outside of the United States. But future American leaders should look toward Winston Churchill and Abraham Lincoln rather than a Neville Chamberlain for their role models. Leadership can be lonely, and it may be a generation or more before such first-strike military actions will be recognized as prudent. That means America will require political leadership that is less focused on polls, and more focused on governance.

The use of military force is also valid *if our national or economic interests are threatened.* Three primary threats to America's national interest exist in the 21st century: (1) the interruption in the flow of vital natural resources such as oil or coal, iron, or other strategic minerals; (2) a denial of access to the world's critical maritime trade routes; (3) or an act of aggression against a U.S. ally or trading partner.

A catastrophic interruption in the distribution of strategic materials would be devastating to both the national and international economy. The United States is the powerhouse of the world's economic system, and the world's number-one consumer of foreign goods. The people of 100 nations rely on a strong America to fuel their own businesses, their own economies. As the world's greatest generator of wealth, the United States must retain a healthy, vibrant, and growing economy.

Despite the cycle of economic highs and lows, of booms and busts, many nations of the world have evolved economically in the 50 years since the end of the last world war. Whole populations—in Taiwan, South Korea, Japan, Mexico, India, Pakistan, the Philippines, and Indonesia—have moved from the brink of poverty and want, to lifestyles comparable to the middle classes in Europe and America. This economic growth and progress propels positive social evolution, resulting in greater opportunity for personal growth, freedom of religion, and gender equality. In other words, personal affluence and civic stability lead to positive human evolution, so such economic stability and growth must continue if humans as a species are to survive into the 22nd century.

Bad economies have negative consequences, and the economic and political chaos isn't exclusive to emerging nations. Long-established nation-states can also fall apart and reemerge in dangerous new ways in times of economic stress. During the political chaos following Germany's defeat in World War I, a nation that once prided itself on its progressive principals found the streets of its cities held hostage by Communist and Nazi thugs battling for control of political power. Then came the Great Depression, which further deepened Germany's economic hardship. More political turmoil followed. To escape their plight, the German people did what the populations of emerging states do today—they found a despot to lead them, a scapegoat in an ethnic and religious minority that they quickly oppressed, and then they extended their aggression into neighboring Poland and Czechoslovakia—and eventually into France, Belgium, Holland, Eastern Europe, and the Soviet Union.

Now that we know that people in desperate economic straits (even people from the one of the most civilized nations in the world) will do desperate things in desperate times, it is up to governments of the West—and the United States of America in particular—to do everything possible to guarantee that such economic hardship will never again become widespread or endemic, at least in the post-industrial Western democracies and the democracies in East Asia. Only a stable civilization can pull the less civilized nation-states out of the mire of primitivism.

The denial of access to trade routes would also have a debilitating effect on America's national and economic interests. There are critical choke points where commercial shipping can be halted—the Panama or Suez Canals, for instance, or in the Persian Gulf or the Malaysian Straits. It is for that reason that the United States has taken on Great Britain's role of maintaining the most powerful naval force on the face of the earth. America's ability to project military power anywhere in the globe in a very short time assures that crucial maritime trade routes will be open for all the peaceful nations of the world.

If one of America's critical trade partners, such Japan or South Korea, were to be attacked, the ripple effect on the U.S. economy would quickly become a tidal wave. In a global economy, any interruption in the natural flow of trade and commerce can be dangerous. But when a close ally and long-time trading partner is threatened, the danger is greatly magnified. That is why the United States still maintains thousands of troops in South Korea, in Japan, and in Europe. It is wise and prudent to protect one's closest allies from outside aggression.

Since the fall of the Soviet Union, the threat of war in Europe has been greatly diminished, along with the strategic importance of NATO. But the threat in Korea is, if anything, more ominous than before. And America's long-time allegiance to the state of Israel—still the only democratic state in the Middle East—and Taiwan also assures us future political grief and possible aggression. Our strategic alliances may yet force the U.S. government into a military confrontation in those troubled regions.

Humanitarian aid and disaster relief can be valid reasons to deploy U.S. troops, and this notion has appealed to Americans in the late 20th century as a means to help the world's unfortunates. American forces have often been dispatched to keep order and to assist during such crises, and modern military efforts to help the poor can be traced back to the U.S. Army Special Forces' Special Proficiency at Rugged Training and Nation-Building (SPARTAN) program—a domestic improvement social program started in the 1970s.

Internationally, America has interceded a half-dozen times in the last few decades to provide aid and comfort to suffering people. If such missions are performed through the auspices of the United Nations, or after a coalition of developed nations have reached a consensus on what actions should be taken, then humanitarian operations can have a beneficial effect—but only if the troops deployed are perceived by the recipients of the aid as deliverers, not as a hostile occupation force.

This lesson was best demonstrated in Somalia. The U.S. Marines did a fine job of ending the immediate threat of starvation for a half-million Somalis in 1991 and 1992, when the first President Bush deployed them. But when the U.S. Army Rangers and Delta Force replaced the Marines in 1993, and immediately tried to round up the corrupt warlords who caused the suffering, the American troops started to look more like a foreign occupational force to the Somalis. It didn't matter that the Army was earnestly trying to end the root causes of famine, rather than temporarily alleviate the suffering.

In the end, half the city of Mogadishu rose up against them. Somali fury, coupled with a U.S. administration that forbade the deployment of U.S. armor and AC-130 Specter gun ships, led to an American military disaster. The lesson for future military deployment for humanitarian aid becomes clear: Next time bring the carrot *and* the stick.

Halting genocide. The word genocide comes from the Greek *genos*, which means *people* or *race*, and from the Latin *caedere*, which means *to kill*. The word was coined in 1944 by Polish writer Raphael Lemkin to describe the Nazi atrocities in Eastern Europe. Though the term origi-nated in the 20th century, the concept and practice of genocide is far older.

Genocidal practices have been all too common throughout history, from the beginning of the Christian Crusades in the eleventh century through the brutality of Genghis Khan, to Timur's brutal practices in Persia, India, and Syria in the 14th century, to the wars of European Reformation, in which organized massacres of religious opponents was openly practiced. More modern examples include the near eradication of the Northeastern American Indian tribes by the first European set-tlers, the massacre of Armenians at the hands of the Turks in 1915, and Josef Stalin's violent suppression of the kulaks of Eastern Europe. Though the brutality of genocide reached a new level of technological efficiency under the Nazis, Hitler and company hardly invented the concept.

Today, genocide is defined by the 19 articles that make up the Geno-cide Convention that came into force in January 1951. Today, partici-pating in genocide or genocidal events is a punishable offense under international law. Of course, questions arise in the enforcement of such laws—especially questions about the possible violation of state sovereignty—though, since the internationally sponsored intervention in Kosovo in 1999, it is now generally accepted that genocide is not an internal domestic matter, but is a problem to be dealt with through independent, unbiased, international intervention.

Yet as horrible, ghastly, and unspeakably inhuman as genocide is, it is not always in the best interests of the United States to become involved in the genocidal activities of other nations. As much as we want to end genocide on a moral level, with no direct threat to American hegemony, U.S. administrations have recognized that to get involved in every occurrence puts too great a strain on our finite military resources. Though the moral justification to halt genocide is always present— the same question always arises: Where do we, as Americans, start to eradicate genocidal activities? In the Sudan or the Philippines, where Muslim extremists enslave, murder, and sometimes behead Christians? Or the Middle East, where Palestinians have vowed to destroy Israel even as the Israelis grab land and level their houses? Do we become involved in Kosovo, where Serbian Christians are oppressing their Muslim citizens? Should we have gone into Cambodia, to halt the Communist Khmer Rouge massacres, or Indonesia to halt the massacre of Communists? Should we have attacked Communist China during its bloody Cultural Revolution? Or Stalin's Russia during the forced collectivization of kulak farms? Or the Hutus? The Tutsis?

Too often the problem of genocide cannot be solved through military action anyway. It takes much more: An international consensus is required, along with the political will to prevail against state-sponsored genocide, no matter what the cost. Indeed, the world's response to outbreaks of genocide since World War II has generally been tepid. At various times there has been great public outcry—about the killing of Serbs in Kosovo in 1999, for instance. But many more acts of racial, religious, or ethnic genocide have gone unpunished and unnoticed by the international community. What nation took action to end the brutality of the Khmer Rouge in Cambodia? Or the mass killings of the Ibo that sparked the Nigerian Civil War? Or for Christians who are— right now—being butchered and enslaved in present-day Sudan by their Muslim masters? The world seems to have a selective blindness when it comes to genocide. Some peoples are lucky, and the world collectively rushes to their aid. But woe to any culture, tribe, or religion that fails to gain the world's attention.

Curtailing nuclear proliferation has emerged most recently as a justification for military action, and it is a valid one. Nuclear proliferation is defined as the process by which one nation after another comes into possession of, or the right to determine the use of, nuclear weapons for offensive or defensive military action.

Although the concept of stopping the spread of nuclear weapons once the genie has been unleashed from his bottle might seem absurd, there are many sane reasons to stop weapons of mass destruction from falling into the "wrong hands"—especially if those hands belong to nation-states with erratic military or governmental leadership, a penchant for genocide, or close ties with ruthless, stateless terrorist or fundamentalist religious groups that bear grudges against civilization in general and the Western democracies in particular.

Nations with a stake in civil order and international stability may be trusted to use nuclear weapons in the same way the Western democracies have used them—not at all in half a century. However, rogue nations with no stake in the future of humanity, suicidal religious tendencies, or a visceral hatred of their neighbors should be prevented from obtaining weapons of mass destruction at all costs—including the total annihilation of a rogue state or two, if necessary. This is not colonialism. This is not "keeping the Third World down." This is the prevention of a nuclear exchange and its potential for regional or even global destruction.

As in the case of genocide, the democracies of the West can do only so much to prevent nuclear proliferation. But all of humanity has a vested interest in stopping certain rogue nations and belligerent governments from obtaining such weapons. The trick lies in realizing which nations present the most immediate threat, and which nations are simply trying to join the less and less exclusive club of nuclear powers for their own self-defense. We do not (necessarily) quake with fear if Turkey tests a nuclear weapon, but no sane person would want Zimbabwe despot Robert Mugabi to have a single nuclear warhead.

Of course, nuclear weapons themselves are only part of the problem. A nation can possess nuclear weapons but lack an effective delivery system with which to deploy them. China, for instance, has been a nuclear power since at least the 1980s, but until they obtained advanced guidance microchips from a U.S. defense contractor in the 1990s, they were still years away from developing an effective intercontinental ballistic missile. As of this writing, the Chinese are perhaps a year or so away from testing a missile capable of striking anywhere in the continental United States. If America's relationship with the Chinese government cools, another arms race rivaling the massive size and scope of Soviet-American build-up may occur. In such circumstances, the United States will have no choice but to aggressively pursue the feasibility of the Strategic Defense Initiative—the so-called Star Wars defense program—to its logical or illogical conclusion.

War is a risky business. You can hold all the cards, or can think you do, but once the first shots are fired, surprising and chaotic things begin to happen. Things fall apart, the center does not hold, or, at the very least, events move so quickly that they can easily spiral out of control.

So any leader embarking on a path to war had better have a very good reason—and the force to back it up. He or she must be capable of arguing his or her case before the democratic people of the West, and form some sort of consensus. There are good and honorable reasons to wage war: in the event of attack, to prevent terrorism, to protect the national interest when it is threatened, to protect an ally when it is attacked, to halt genocide, or to curtail the proliferation of nuclear weapons—but war, by its very nature, is never good.

NO NUKES IS GOOD NUKES

All through the Cold War, as the United States and the Soviet Union built their stockpiles of nuclear weapons, they also negotiated to limit the use and deployment of such weapons. During the Nixon administration the first Strategic Arms Limitations Talks (SALT) were conducted. The Anti-Ballistic Missile Treaty of 1972 somewhat limited nuclear weapons. Additional terms for SALT II were negotiated by Jimmy Carter, but when the Soviet Union invaded Afghanistan the treaty was abandoned.

Ronald Reagan resisted negotiations, and his deployment of intermediate-range nuclear missiles in Europe alarmed the Soviets. Premier Mikhail Gorbachev urged the Americans back to the bargaining table and the Intermediate Nuclear Missile Treaty of 1987 represented the first true reduction in nuclear weapons by both countries.

The START Treaty of 1991 reduced long-range nuclear weapons, and was signed before the collapse of the Soviet Union. Presidents George Bush and Boris Yeltsin further reduced the number of nuclear weapons by signing the SALT II treaty over a decade after the agreement was proposed, but neither the Russian Duma nor the U.S. Congress have ratified it. President Clinton signed the Comprehensive Test Ban Treaty in 1996, which would have banned all future nuclear weapons testing, but Congress refused to ratify it, too. President George W. Bush has since pulled out of the SALT II agreement, and is unwilling to sign the Comprehensive Test Ban even though nuclear tests have mostly been limited to computer models since the mid-1990s.

The best way to win a war is to not get involved in one in the first place. But if war is inevitable, then victory is better than defeat. How can victory be achieved in our complex, post-industrial age of global economies and international politics?

Historically, Western civilization has defined victory in different ways. Total victory is defined as the annihilation of the enemy's military force and the overthrow of their government, but not necessarily the destruction of the enemy's culture and society. Ancient societies showed no such qualms—to the Romans, the Spartans, the Mongols, war was absolute and ended only when all members of the opposition were annihilated or sold into slavery, their cities leveled, and salt sown into their earth so that no crops would thrive.

In the 21st century we are, of course, more civilized, and pursue what we think to be civilized policies of warfare. We prefer, for instance, to level an entire city rather than to assassinate the single despot who challenged international order and most probably oppressed his own people to achieve power in the first place. But at least the democracies of the West have moved beyond the need for the total extermination of its foes.

Victory is one of the most basic, yet most elusive concepts in military philosophy. Any time you deny the enemy his goals, or achieve your own objectives, it can be considered a victory. But in real life—and realpolitik—things are seldom so clear cut.

From the 17th to the dawn of the 20th century, victory in European war was achieved in two separate and distinct phases: first, a limited military operation, followed by diplomatic negotiations to end hostilities and achieve a lasting peace. These negotiations sometimes took many years before a formal treaty could be hammered out, but once negotiations were entered upon, hostilities were seldom renewed.

This practice was initially extended to the New World. The American Revolutionary War ended with the Treaty of Paris in 1783, even though all major military operations had ceased two years before. But two wars fought in the Americas—the Civil War and the Indian Wars—evolved into total wars of annihilation, redefining the concept of victory in war for 19th-century Americans.

The American Civil War began as a conflict with limited goals, but as the war dragged on, unconditional surrender became the mantra on both sides. Abraham Lincoln, along with Generals Grant and Sherman, would settle for nothing less than reunion and the end of slavery. The Confederacy wanted nothing less than succession. Only after Grant and Sherman annihilated the South did the war come to an end.

The Indian Wars began with limited goals, but in the end, Native Americans were forcibly relocated, or even exterminated by the superior firepower and mobility of the U.S. Calvary.

During World War I, U.S. president Woodrow Wilson argued against the imposition of a settlement by the victors upon the vanquished, maintaining that such an outcome would only breed resentment and would undermine the prospects for a lasting peace. Wilson proposed a new vision of a peace based upon the interests of all nations and the repudiation of traditional power politics. The war ended in an Armistice—another European-style-negotiated settlement. But the Treaty of Versailles imposed harsh punishment on vanquished Germany, and President Wilson, without membership in the faltering League of Nations, could do nothing to curtail France's vengeful spirit. The result was the very resentment Wilson warned against, the rise of Nazism, and a Second World War.

This tragic blunder shaped American views about victory during World War II, and although both Nazi Germany and Imperial Japan were practically annihilated by the end of hostilities, the unique cultures, society, and citizens of those formerly belligerent nations were permitted to survive and thrive under a more "Wilson-esque" model of international power politics. Through the Marshall Plan, Germany and Japan were provided with the means to rebuild their nation-states, infrastructure, and industrial base in order to compete in world markets as equal citizens and allies of the West, once a period of benevolent occupation had ended. Less fortunate were the nations of Eastern Europe, who fell under the oppressive domination of the Soviet Union, the modern equivalent of a "dark age" from which they are only now beginning to emerge.

During the Cold War, the U.S. government confronted a new strategic environment. The concept of achieving the unconditional surrender of the Soviet Union seemed incomprehensible, so a rising class of civilian strategists developed theories of "limited war," and President Harry Truman developed the "Truman Doctrine" of containment of Communism, rather than the annihilation of the Communist states through military confrontation. In this climate, the Korean War was seen as the new norm—a limited war fought to stop Communist advances, which ended with the restoration of the *status quo*—the division of Korea into a free, capitalist South Korea and an oppressed, Communist North Korea that persist to the present day.

In Korea, the United States accepted *stalemate* and *containment*, rather than extending hostilities any further and risking a prolonged military confrontation with China.

Today, the concept of containment, which began with the Truman Doctrine in 1946, is still considered a viable option in situations where the total annihilation of a nation-state is not possible, or is politically risky. But the problems with containment are many. Without total victory, the war is not yet finished, and the underlying factors that led to war usually persist. In time, those conditions may again become intolerable and the conflict will begin anew. Containment is also expensive. In the case of Korea, thousands of U.S. troops have continued to patrol the borders of North and South Korea today, a half century after hostilities supposedly ended.

The concept of limited war led the Kennedy and Johnson administrations to the "Flexible Response" strategy, first postulated in General Maxwell Taylor's book *The Uncertain Trumpet* (1960). Taylor was the chief of staff of the U.S. Army, until he objected to budget cuts and U.S. strategists' focus on a nuclear response at the expense of conventional forces, and resigned in 1959. In his study of war in the post-nuclear age, General Taylor searched for ways in which nuclear weapons and conventional weapons could be used jointly in battle without sparking an all-out nuclear war. John F. Kennedy was impressed by General Taylor's theories and adopted some of his concepts. Though Flexible Response persists as part of official U.S. military policy, it is being eclipsed in importance by the increasing awareness that nuclear weapons have no tactical use, and are only valuable as a deterrent. In any case, Flexible Response and the concept of Limited War ended with America's military defeat in Vietnam.

This defeat led to a reassessment of the relationship between the military and political factors in war. The result was the Weinberger and Powell Doctrines (named after Secretary of Defense Casper Weinberger and General Colin Powell), implemented first in Panama in 1989 and 1990, and then in the Persian Gulf War of 1991. The tenets of these doctrines revolve around the application of overwhelming military force to terminate a conflict both swiftly and decisively. Though the Weinberger Doctrine worked well enough in Panama, despite the over-whelming military victory against Iraq in the Persian Gulf War, true victory was denied because Saddam Hussein remained in power, and remained a threat to peace in the region. In the years following that con-flict, the United States and its coalition allies have had to resort to con-tainment to prevent Hussein's forces from again crossing their borders, and (as of this writing) another war against Iraq is a real possibility.

As with every other aspect of human conflict, the lethal technologies of modern warfare have forced the nations of the world to reassess the very concept of military victory. Even though the United States entered the 21st century as the world's only superpower, our nation still wrestles with the central problem of achieving the proper balance between polit-ical objectives and military means. Despite the vast arsenal of powerful weapons at our disposal, we still walk a tightrope stretched over an abyss of nuclear or biological annihilation, and each step we take is fraught with peril.

APPENDIX A

RECOMMENDED READING

Hundreds of sources were used in the preparation of this manuscript. Information came from periodicals, newspaper stories, Pentagon press releases, and on- and off-the-record interviews with members of the U.S. military, defense contractors, authors of fiction and nonfiction, and specialists in many fields—far too many to cite here.

However, dozens of websites and books have also supplied source material, and these are worth mentioning.

INTERNET SOURCES

Following is a list of the best Internet sources for further information on the weapons and tactics of future conflict. The list is divided (roughly) into categories.

GENERAL INFORMATION

www.brook.edu/dybdocroot/default.htm
The Brookings Institution
A think tank dealing with the politics, strategy, and tactics in the post–Cold War era

www.fas.org/index.html
Federation of American Scientists
Good general military information, but content often displays an anti-military bias

www.foresight.org/homepage.html
Foresight Institute
Dedicated to the application of emerging nanotechnologies

www.ornl.gov
Oak Ridge National Laboratories

www.comw.org/rma/#
Revolution in Military Affairs
Site provides an open forum to debate strategies and tactics of post–Cold War asymmetrical warfare

INFANTRY AND LAND WARFARE

www.generaldynamics.com
General Dynamics Land Systems
Corporate site

carlisle-www.army.mil/usawc/Parameters/a-index.htm
Parameters: The United States Army War College Quarterly

www.army-technology.com/projects/index.html
Site for Defense Industries—Army
Excellent overview of advanced weapons systems for the U.S. Army

www.army.mil/cmh-pg
U.S. Army Center of Military History

www.natick.army.mil
U.S. Army Soldier Systems Center
Future land warfare technology

www.army.mil
The U.S. Army's official site

U.S. SPECIAL OPERATIONS FORCES

www.sfahg.com
Special Forces Search Engine
Excellent links to other special operations sites

www.specialoperations.com/usspecops.html
Special Operations.com
Information on worldwide special operations and a list of terrorist
organizations/activities

www.governmentguide.com/officials_and_agencies/specialops.adp
U.S. Special Operations Information Page

www.specialoperations.com/Army/Special_Forces/
U.S. Army Special Operations Information Page
Covers the Green Berets exclusively

NAVAL AND AMPHIBIOUS WARFARE

www.maritime.org/search.htm
Maritime Park Association
Good search engine for naval information

www.navsea.navy.mil
Navesea: Naval Sea Systems Command
Site dedicated to the dissemination of information about new naval
technology

www.oecswath.com/default.asp
Ocean Engineering Consultants, Inc.
Excellent source for detailed and highly technical explanations of the
different types of SWATH technologies

www.chinfo.navy.mil/navpalib/factfile/ffiletop.html
United States Navy Fact File
Excellent source of information about U.S. Navy weapons systems

www.usmc.mil
The United States Marine Corps' official site

www.navy.mil
The United States Navy's official site

CHEMICAL, BIOLOGICAL, AND NUCLEAR WARFARE

www.cbiac.apgea.army.mil
Chemical and Biological Defense Information Analysis Center

www.sbccom.army.mil/
U.S. Army Soldier and Biological Chemical Command

AIR WAR

airforce-technology.com/
Air Force Technology
Up-to-date information about new U.S. Air Force research programs

www.airpower.maxwell.af.mil/
Air and Space Power Chronicles
The Department of Defense's first professional online journal

www.aviation-history.com/index.html
Aviation History Online Museum
Useful site for information on the evolution of warplanes

www.f22-raptor.com
F/A-22 Raptor Team Infonet

www.lockheedmartin.com
Lockheed Martin
Corporate site

www.northgrum.com
Northrop Grumman
Corporate site

www.af.mil/sites
The U.S. Air Force official site

WAR IN SPACE

www.peterson.af.mil/hqafspc/default2.asp
Air Force Space Command official site

www.spacewar.com
SpaceWar: Your Portal to Military Space
Up-to-date information about space warfare with good links to other
future war sites

BOOKS

The following is a list of recommended reading, divided into three
categories: "General Reference" offers titles valuable in gaining important background information on military subjects, the study of military
history, or books pertaining to the subject of future war literature. In
"Guides to Specific Training, Weapons, and Technology," you'll find
books that specifically focus on one subject, one weapon system, one
training regimen, one campaign or branch of service, and so on. The
final self-explanatory category is "Strategy, Tactics, and Philosophies of
Future Warfare."

GENERAL REFERENCE

Ambrose, Stephen E. *Band of Brothers*. New York: Simon & Schuster,
    1989.

———. *D-Day*. New York: Simon & Schuster, 1994.

Anderson, Bern. *By Sea and by River: The Naval History of the Civil
    War*. New York: Knopf, 1962.

Atkinson, Rick. *Crusade: The Untold Story of the Persian Gulf War*.
    Boston and New York: Houghton Mifflin, 1993.

Bergerud, Eric M. *Fire in the Sky: The Air War in the South Pacific.* Colorado: Westview Press, 2000.

Bobrick, Benson. *Angel in the Whirlwind: The Triumph of the American Revolution.* New York: Simon & Schuster, 1997.

Bowden, Mark. *Black Hawk Down: A Story of Modern War.* New York: Penguin Books, 1999.

Chambers, John Whiteclay II, editor in chief. *The Oxford Companion to American Military History.* New York: Oxford University Press, 1999.

Clarke, I. F., ed. *The Tale of the Next Great War, 1871–1914.* New York: Syracuse University Press, 1995.

———. *Voices Prophesying War: Future Wars, 1763–3749* (Revised Edition). Oxford: Oxford University Press, 1992.

Corn, Joseph J., and Brian Horrigan. *Yesterday's Tomorrows.* Baltimore: John Hopkins University Press, 1996.

Dugan, James, and Carroll Stewart. *Ploesti: The Great Ground-Air Battle of 1 August 1943* (Revised Edition). Dulles, Virginia: Brassey's, Inc., 2002.

Dunnigan, James F. *How to Make War: A Comprehensive Guide to Modern Warfare* (First Revised Edition). New York: William Morrow, 1988.

Dunnigan, James F., and Albert A. Nofi. *The Pacific War Encyclopedia.* New York: Checkmark Books, 1998.

Dupuy, Trevor N., Curt Johnson, and David L. Bongard. *The Harper Encyclopedia of Military Biography.* New York: HarperCollins Publishing, 1992.

Farwell, Byron. *Encyclopedia of 19th Century Land Warfare.* New York: W.W. Norton & Company, 2001.

Foss, Joe, and Matthew Brennan. *Top Guns: America's Fighter Aces Tell Their Stories.* New York: Pocket Books, 1991.

Holmes, Richard. *The Oxford Companion to Military History*. Oxford: Oxford University Press, 2001.

Kessler, Ronald. *Inside the CIA*. New York: Simon & Schuster, 1992.

Kitz, Janet F. *Shattered City: The Halifax Explosion and the Road to Recovery*. Halifax: Nimbus Publishing Limited, 1989.

Laffin, John. *Brassey's Battles*. London: Brassey, 1986.

Laur, Colonel Timothy M., and Steven L. Llanso. Edited by Walter J. Boyne. *The Army Times, Navy Times, Air Force Times Encyclopedia of Modern U.S. Military Weapons*. New York: Berkley Books, 1995.

Luttwak, Edward, and Stuart L. Koehl. *The Dictionary of Modern War*. New York: HarperCollins, 1991.

Perret, Geoffrey. *A Country Made by War: From the Revolution to Vietnam—the Story of America's Rise to Power*. New York: Random House, 1989.

Proft, R. J., ed. *United States of America (Congressional) Medal of Honor Recipients and Their Official Citations*. Columbia Heights: Highland House II, 1997.

Sakai, Saburo, with Martin Caiden and Fred Saito. *Samurai!* New York: ibooks, 2001.

Tuchman, Barbara W., with Preface by Robert K. Massie. *The Guns of August: The Reprint Edition*. New York: Ballantine, 1994.

Tucker, Spencer C., ed. *The Encyclopedia of the Vietnam War: A Political, Social, and Military History*. New York: Oxford University Press, 2000.

GUIDES TO SPECIFIC TRAINING, WEAPONS, AND TECHNOLOGY

Alexander, Colonel John B. *Future War: Non-Lethal Weapons in 21st-Century Warfare*. New York: St. Martin's Press, 1999.

Beckwith, Colonel Charlie A. (Ret.), with Donald Knox. *DELTA Force: The Army's Elite Counter-Terrorism Unit*. Novato: Presidio, 1983.

Bohrer, David. *America's Special Forces: Weapons, Missions, Training*. Osceola, WI: Motorbooks International, 1998.

Caidin, Martin. *The B-17: The Flying Forts*. New York: ibooks, 2001.

Campbell, R. Thomas. *The C.S.S. Hunley: Confederate Submarine*. Shippensburg, PA: Bard Street Press, 2000.

Canfield, Bruce N. *U.S. Infantry Weapons of World War II*. Rhode Island: Andrew Mowbray, 1994.

Chant, Chris. *German Warplanes of World War II*. London: Amber Books, 1999.

Chapelle, Howard I. *The History of the American Sailing Navy: The Ships and Their Development*. New York: W.W. Norton, 1949.

Clancy, Tom, with John D. Gresham. *Airborne: A Guided Tour of an Airborne Task Force*. New York: Berkley Books, 1997.

————. *Special Forces: A Guided Tour of U.S. Army Special Forces*. New York: Berkley Books, 2001.

Cross, Wilbur. *Zeppelins of World War I*. New York: Barnes & Noble Books, 1993.

de Kay, James Tertius. *Monitor: The Story of the Legendary Civil War Ironclad and the Man Whose Invention Changed the Course of History*. New York: Walker & Company, 1997.

Donald, David, ed. *American Warplanes of World War II*. London: Aerospace Publishing Ltd., 1995.

Dorr, Robert F. *Desert Storm Air War*. Osceola, WI: Motor Books International, 1991.

Ford, Roger. *The World's Great Tanks: From 1916 to the Present Day*. London: Brown Books, 1997.

Franks, Norman. *Aircraft vs. Aircraft*. London: Grub Street, 1998.

Gann, Ernest K. *The Black Watch: The Men Who Fly America's Secret Spy Planes*. New York: Random House, 1989.

Giangreco, D. M. *Stealth Fighter Pilot*. Osceola, WI: Motorbooks International, 1993.

Greene, Jack, and Alessandro Massignani. *Ironclads at War: The Origin and Development of the Armored Warship, 1854–1891*. Conchohocken, PA: Combined Publishing, 1998.

Griswald, Terry, and D. M. Giangreco. *Delta*. Osceola, WI: Motorbooks International, 1992.

Gunston, William. *Combat Arms: Modern Helicopters*. New York: Prentice Hall Press, 1990.

Halberstadt, Hans. *Desert Storm Ground War*. Osceola, WI: Motor Books International, 1991.

———. *U.S. Navy SEALs*. Osceola, WI: Motor Books International, 1993.

Hart, Dr. S., and Dr. R. Hart. *German Tanks of World War II*. New York: Doubleday, 1998.

Hinchliffe, Peter. *The Other Battle: Luftwaffa Night Aces versus Bomber Command*. Shrewsbury, UK: Airlife Publishing Ltd., 1996.

Longstreet, Stephen. *The Canvas Falcons*. New York: 1970.

Macksey, Kenneth. *Tank vs. Tank*. London: Grub Street, 1988.

Marquis, Susan L. *Unconventional Warfare: Rebuilding U.S. Special Operations Forces* (The Rediscovering Government Series). Washington, D.C.: Brookings Institute, 1997.

Marshall, Chris, General Editor. *The Encyclopedia of Ships*. London: Brown Books, 1995.

McRaven, William H. *Spec Ops: Case Studies in Special Operations Warfare, Theory and Practice*. Novato: Presidio Press, (Reissue) 1998.

Mindell, David A. *War, Technology, and Experience Aboard the USS Monitor*. Baltimore and London: John Hopkins University Press, 2000.

Oliver, David, and Mike Ryan. *Warplanes of the Future*. London: Salamander Books, 2000.

Pushies, Fred J. *U.S. Air Force Special Ops*. Osceola, WI: Motor Books International, 2001.

———. *U.S. Army Special Forces*. Osceola, WI: Motor Books International, 2000.

———. *Weapons of Delta Force*. Osceola, WI: Motor Books International, 2002.

Rich, Ben R., and Leo Janos. *Skunk Works*. New York: Little, Brown and Company, 1994.

Richardson, Doug. *Stealth Warplanes*. Osceola, WI: Motor Books International, 2001.

Rosen, Stephen Peter. *Winning the Next War: Innovation and the Modern Military*. Ithaca, NY: Cornell University Press, 1994.

Spate, Wolfgang. *Top Secret Bird: The Luftwaffa's Me-163 Comet*. Missoula, MT: Pictorial Histories Publishing Co., 1989.

Sweetman, Bill. *F-117 Nighthawk*. St. Paul, MN: MBI Publishing Company, 1995.

———. *Inside the Stealth Bomber*. St. Paul, MN: MBI Publishing Company, 2000.

———. *Lockheed Stealth*. St. Paul, MN: MBI Publishing Company, 2001.

Terraine, John. *The U-Boat Wars: 1916–1945*. New York: G.P. Putnam's, 1989.

Wright, Patrick. *Tank*. New York: Viking, 2002.

Yoshimura, Akira, translated by Retsu Kaiho and Michael Gregson. *Zero Fighter*. Westport: Praeger, 1996.

STRATEGY, TACTICS, AND PHILOSOPHIES OF FUTURE WARFARE

Friedman, George, and Meredith Friedman. *The Future of War: Power, Technology, and American World Dominance in the Twenty-First Century*. New York: St. Martin's Griffin, 1996.

O'Hanlon, Michael E. *Technological Change and the Future of Warfare*. Washington, D.C.: The Brookings Institution Press, 2000.

Peters, Ralph. *Beyond Terror*. Mechanicsburg, PA: Stackpole Books, 2002.

———. *Fighting for the Future*. Mechanicsburg, PA: Stackpole Books, 2001.

Skolnikoff, Eugene B. *The Elusive Transformation*. Princeton, NJ: Princeton University Press, 1994.

APPENDIX B

NOTES

CHAPTER 4

[1] Oak Ridge Laboratory Status Report, February 26, 2002.

[2] Soldier Systems Center press release, 2002.

CHAPTER 5

[1] Friedman, George, and Friedman, Meredith. *The Future of War: Power, Technology and American World Dominance in the Twenty-First Century.* New York: St. Martin's, 1998, p. 393.

CHAPTER 7

[1] Rich, Ben R., and Leo Janos. *Skunk Works: A Personal Memoir of My Years at Lockheed.* New York: Little, Brown, 1994.

[2] Ibid.

CHAPTER 8

[1] Clancy, Tom. *Marine: A Guided Tour of a Marine Expeditionary Unit.* New York: Berkley Books, 1996.

CHAPTER 9

[1] Richardson, Doug. *Stealth Warplanes*. Osceola, WI: Motor Books International, 2001.

[2] Sweetman, Bill. *Lockheed Stealth*. Osceola, WI: Motor Books International, 2001.

[3] Kaczor, Bill. "Drones May Be in Air Force Future." Associated Press. June 23, 2002.

CHAPTER 10

[1] U.S. Government Procurements, *Commerce Business Daily*, no. PSA-0547. March 9, 1992.

INDEX

H

J–K

L

O–P–Q

ABOUT THE AUTHOR

Marc Cerasini is the author of *Heroes: U.S. Marine Corps Medal of Honor, The Complete Idiot's Guide to U.S. Special Ops Forces, The New York Times* bestseller *American Hero, American Tragedy,* and the *USA Today* bestselling children's biography *Diana, Queen of Hearts.* He has also written several books of popular culture, including *24: The House Special Subcommittee's Findings at CTU,* a novel based on the Emmy Award–winning Fox Network television drama, and the lead essay on the history of the modern techno-thriller found in *The Tom Clancy Companion.*